Ana Kapaj

Resource Use Efficiency of Olive oil production in Albania

AF154038

Ana Kapaj

Resource Use Efficiency of Olive oil production in Albania

LAP LAMBERT Academic Publishing

Impressum / Imprint

Bibliografische Information der Deutschen Nationalbibliothek: Die Deutsche Nationalbibliothek verzeichnet diese Publikation in der Deutschen Nationalbibliografie; detaillierte bibliografische Daten sind im Internet über http://dnb.d-nb.de abrufbar.
Alle in diesem Buch genannten Marken und Produktnamen unterliegen warenzeichen-, marken- oder patentrechtlichem Schutz bzw. sind Warenzeichen oder eingetragene Warenzeichen der jeweiligen Inhaber. Die Wiedergabe von Marken, Produktnamen, Gebrauchsnamen, Handelsnamen, Warenbezeichnungen u.s.w. in diesem Werk berechtigt auch ohne besondere Kennzeichnung nicht zu der Annahme, dass solche Namen im Sinne der Warenzeichen- und Markenschutzgesetzgebung als frei zu betrachten wären und daher von jedermann benutzt werden dürften.

Bibliographic information published by the Deutsche Nationalbibliothek: The Deutsche Nationalbibliothek lists this publication in the Deutsche Nationalbibliografie; detailed bibliographic data are available in the Internet at http://dnb.d-nb.de.
Any brand names and product names mentioned in this book are subject to trademark, brand or patent protection and are trademarks or registered trademarks of their respective holders. The use of brand names, product names, common names, trade names, product descriptions etc. even without a particular marking in this works is in no way to be construed to mean that such names may be regarded as unrestricted in respect of trademark and brand protection legislation and could thus be used by anyone.

Coverbild / Cover image: www.ingimage.com

Verlag / Publisher:
LAP LAMBERT Academic Publishing
ist ein Imprint der / is a trademark of
OmniScriptum GmbH & Co. KG
Heinrich-Böcking-Str. 6-8, 66121 Saarbrücken, Deutschland / Germany
Email: info@lap-publishing.com

Herstellung: siehe letzte Seite /
Printed at: see last page
ISBN: 978-3-659-26515-0

Zugl. / Approved by: Faculty of Agricultural Sciences, 2014

Resource Use Efficiency of Olive oil production in Albania

by

Ana Kapaj, PhD

(Post-Doctoral Study)

Acknowledgements

I would like to express my very deep thanks to the Islamic Development Bank (IDB) which has supported this study as part of my Post-Doctoral research.
THANK YOU

Dedication

This study is dedicated to my FAMILY, my very dear husband and my two lovely daughters, for their demonstration of love, commitment and support during my absence. You have made more of an impression on me than you know.

List of Contents

List of Figures

List of Tables

List of Abbreviations

APC:	Agricultural Production Cooperative
CEEC:	Central Eastern Europe Countries
C.I.F.:	Cost, Insurance and Fright
DRC:	Domestic Resource Cost
EU:	European Union
F.O.B.:	Free on Board
FTER:	Free Trade Exchange Rate
GDP:	Gross Domestic Product
IFDC:	International Fertilizer Development Centre
INSTAT:	Institute of Statistics
ISMEA:	Istituto di Servizi per il Mercato Agricolo Alimentare
MAF:	Ministry of Agriculture and Food
OER:	Official Exchange Rate
PAM:	Policy Analysis Matrix
PCR:	Private Cost Ratio
PPC:	Production Possibility Curve
SCF:	Standard (or social) Conversion Factor
SER:	Shadow Exchange Rate
SF:	State Farms
UNCTAD:	United Nations Conference on Trade and Development
WTO:	World Trade Organization

PART I

1. Introduction

1.1. Problem statement and research objectives

Following the drastic changes after the year 1990, the Albanian agriculture encountered numerous problems that caused considerable changes in the market structure. The liberalisation of the trade system started right after this year. Now, Albania applies an open and liberalized trade system (according to the decision of the Council of Ministers, dated 16.09.1999). During these 24 years, Albania has gone through a series of political and socio-economic changes which transformed the agricultural production and agro-industry of the country. The new situation created has its impact on the balance of the foreign trade for agricultural products. Imports of these goods increased substantially followed by a considerable reduction of exports in these products.

Like a lot of other products, the major part of the oil (vegetal and olive) in Albania comes from import. This is because of the difficulties in insuring the raw material needed for the processing oil industry and the absence of the infrastructure and the adapt industry. The farmers are not interested in oil plants because of the low economic profitability and furthermore the already existing plants were destroyed after the 1990s. If Albania will reach an average yield compared to that of the neighbour countries (Greece and Italy), then the olive would turn into an important resource for the oil industry (MoAFCP, 2012).

This study aims at assessing resource use efficiency of the olive oil manufacturing industry, evaluating the marketing potential of olive oil, identifying the comparative advantages of olive oil production in Albania.

1.2 Hypotheses and methodology

The purpose of this study is to assess the olive oil production in Albania with the purpose to identify the existence of comparative advantage for olive oil production. The hypothesis to be tested is:

Albania has a comparative advantage in olive oil production under the current economic conditions.

For the assessment of the situation of olive oil production in Albania, a descriptive analysis was used, and for the determination of the comparative advantage of olive oil production the Domestic Resource Cost (DRC) methodology was applied.

The survey[1] was conducted for the sample of 80 processing plants out of 126 actually operating plants in Albania. The central and south-western parts of the country are selected as study area.

Primary and secondary data were used for reaching the objectives. The primary data come from questionnaires that are filled up in 80 olive oil processing plants. For the selection of the 80 processing plants a random sampling was thought to be the best way to guarantee a representative sample. A face to face interview was done in all the selected plants. Using mailed questionnaires in the case of Albania is impossible due to the lack of network and communications skills via e-mail. The survey was carried out from April to June 2013 and the data were collected for the production year 2012-2013. The questionnaire was the product of literature review and discussion with people with experience in the field of questionnaire design. They were constructed in a way to get qualitative and quantitative data.

The questionnaires were the base for constructing the budget of the processing plants. Secondary data comes from the Ministry of Food and Agriculture (MFA), the Institute of Statistics (INSTAT) and International Fertilizer Development Centre (IFDC) centre in Tirana.

---

[1] The term "survey" is commonly applied to a research methodology designed to collect data from a specific population, or a sample from that population, and typically utilises questionnaires as a survey instrument (Robson, 1993).

2. An overlook on the geographical position and Albanian's economy

2.1. Geographical position of Albania

Albania is situated on the eastern shore of the Adriatic Sea, with Montenegro and Serbia to the north, Macedonia to the east, and Greece to the south. Albania may be divided into two major regions: a mountainous highland region (north, east, and south) constituting 70% of the land area, and a western coastal lowland region that contains nearly all of the country's agricultural lands and is the most densely populated part of Albania.

Figure 2.1: Map of Albania
Source: www.shqiperia.com (2014)

Albania has a total surface area of 28.750 square km and a population of 3.544.808 (INSTAT, 2013). Due to the mountains landscape and especially because of its many divisions, the climate varies from region to region. It is warmer in the western part of the country which is mainly under the condition of the warm air masses from the sea (the Adriatic costal region has a typical Mediterranean climate). In the eastern part, which is mostly under the influence of the continental air masses, the winter is cold.

2.2. Macroeconomic situation

Over the past decade, Albania has been one of the fastest-growing economies in Europe, enjoying average annual real growth rates of 6 percent, accompanied by rapid reductions in poverty. Between 2002 and 2008, poverty in the country fell by half (to about 12.4 percent) and extreme poverty now affects less than 2 percent of the population. From 2007 to 2012, the Albanian economy grew by 22%, while exports doubled from 2009 to 2011. Albania has generally been able to maintain positive growth rates and financial stability, despite the ongoing economic crisis. Although Albania has been weathering the impact of the global financial crisis rather well, the recovery to above 3 percent growth rates during 2011 moderated in 2012, reflecting the deteriorating situation in the Eurozone, and the difficult situation in the energy sector. Real GDP growth turned negative in the first quarter of 2012 for the first time since the 2009 crisis. Growth picked up in the second quarter but expectations remain weak for the second half of the year.

Albania's labor market has undergone dramatic shifts over the last decade, contributing to productivity growth. Formal non-agricultural employment in the private sector more than doubled between 1999 and 2011, fuelled largely by foreign investment. Emigration and urbanization brought a structural shift away from agriculture and toward industry and service, allowing the economy to begin producing a variety of services, ranging from banking to telecommunications and tourism.

Despite this shift, agriculture remains one of the largest and most important sectors in Albania. Agriculture is a main source of employment and income – especially in the country's rural areas – and contributes around 17 percent to GDP while accounting for about half of total employment. Albania's agricultural sector continues to face a number of challenges, however, including small farm size and land fragmentation, poor infrastructure, market limitations, limited access to credit and grants and inadequate rural institutions.

Looking toward the future, Albania is focused on supporting economic recovery and growth in a difficult external environment, broadening and sustaining the country's social gains and reducing vulnerability to climate change – particularly through improved water resource management. Key challenges for Albania going forward include early resumption of fiscal consolidation and strengthened public expenditure management, regulatory and institutional reform, reduction of infrastructure deficits, and improvement in the effectiveness of social protection systems and key health services.

Agriculture is one of the most determinative sectors of the Albanian national economy. Its contribution has been decreasing over years and it is estimated at 21% of the GDP in 2013. The rural families continue to dominate the national economy, more than 50 percent of the population lives in the rural areas, and agriculture is the main working alternative of people living in these areas. The real mean growth rate of agriculture production during the last five years is estimated to about 4 percent per year.

Agriculture provides the income basis for most of the population and serves as an employment safety net. The rural population is estimated to comprise about 50 percent of the total population while about 60 percent of the labor force works in agriculture and related fields. Approximately 3250 kinds of plants or 29 % of the species of the European flora and 47 % of the Balkan flora vegetates in Albania.

The agricultural sector suffers from the small size of farms and the fragmentation of farm land, which is a barrier to production and marketing. Higher competitiveness, as a result of lower costs and higher quality, food safety and standards, will strengthen the position of farmers in the market, will raise their income and will introduce safer products in the market for farmers. This is the result of the specific problems that this sector is facing, among which the most evident are the migration from rural areas, land ownership and very limited size of farms, the marketing of products, the irrigation and the drainage system, the low level of technologies in use, the weak organization of farmers, the low development level of agro – processing, etc.

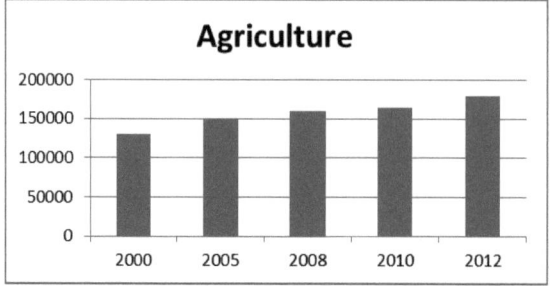

Figure 2.2: Value of agriculture production (Mln ALL)
Source: Ministry of Agriculture, 2013

2.4. Reforms in Agriculture

2.4.1. The reforms

The Land Law of July 1991 specified that the land of Agricultural Production Cooperative (APC) was to be distributed for free and on an equal per capita basis to member families and other rural residents. The privatization of APC proceeded very rapidly: almost 87 percent of collective farm land was distributed by October 1992 (Cungu, Swinnen, 1998). However, the process of land titling was much slower than the actual distribution of land. A number of other decrees on the distribution of other assets and the liquidation of the APC supplemented the Land Law. Livestock, orchards, fruit-trees, etc., were also distributed to farm workers.

In the State Farms (SF), the process differed somewhat from the APC. Only 50,000 hectares of SF lands were initially distributed among farm workers. Most of the remainder was pooled into joint ventures with foreign capital. According to official statistics, by August 1993, 78.8 percent of SF land had been distributed to former specialists and workers, or pooled into joint ventures (Ministry of Agriculture and Food, 1994). When many of these partnerships failed, the land was given to workers for cultivation with the state retaining ownership. In 1995, new legislation stipulated that "use rights" could be turned into "full ownership rights" on ex-joint venture land where disputes over the termination of joint-venture contracts had been cleared.

Initially, land transactions were prohibited. Furthermore, in assigning land ownership or use, the Land Law did not recognize prior land ownership, its pre-collectivization size and boundaries. This caused strong opposition from the pre-1945 owners. They formed a group called the "Property with Justice" movement to increase their influence on political parties. In 1993, the government approved a law on the financial compensation of former landowners but the "Property with Justice" movement rejected the idea of compensation and insisted on full and physical restitution of properties. Although this increased pressure on the government, the Land Law was not fundamentally changed. Instead, in 1995, a new law was passed which provides for ex-owners to receive physical compensation in seaside and tourist site properties.

2.4.2. The reform outcomes

The reforms in agriculture resulted in: a) a fully private production system; b) a complete break up of the APC and SF; c) fragmentation of land ownership and use; and d) strong growth in agricultural output since 1992 (Cungu, Swinnen, 1998).

Despite the incomplete distribution of ownership titles, due to implementation problems, virtually all agricultural production in Albania is in private hands now. All APC-s and most of the SF-s have been completely broken up into individual farms. This process has caused an important fragmentation of land use. More than 95 percent of land is used by around 490.000 individual private farms in at least 1,9 million separate parcels, with an average of about 3,3 separately located parcels for each farm. The average farm size is 1,0 hectare, ranging between 1,3 hectares in valley and foothill regions to 0,8 hectares in the mountains (Cungu, Swinnen, 1998).

Privatization and decollectivization caused considerable changes in incentive structures and in resource allocation. Agricultural output responded by increasing at an average annual rate of 10% for several consecutive years since the reform. This constitutes by far the largest growth rate among all Central Eastern Europe Countries.

3. Olive and olive oil production

3.1 Olive production

3.1.1 Olive cultivation in Albania

Throughout history the olive tree has been a symbol of abundance, glory, and peace. For years olive cultivation has had a very important role in the Mediterranean culture. Olive oil has been more than a mere food to the people of the region; it supposedly has medicinal and magical powers. It is an endless source of fascination and wonder and the fountain of great wealth and power. According to the International Committee of the Olive Oil, more than 80% of the global olive oil production is in Europe and almost 98% of the olives grow in the Mediterranean region.

Albania is one of the few countries in Europe, and the only country in the Central-East Europe, that has the favourable climatic and geographical conditions for olive cultivation. The olive cultivation story in Albania is as old as in the other Mediterranean countries. In the complex process of transition in the Albanian agriculture, the olive and olive oil production will be one of the main directions in the agro food industry. Some of the main reasons that will sustain the potential contribution of the olive sector in the development of the country's economy are:

According to its favourable geographic and climatic conditions, Albania is one of the few countries in Europe where olive can be widely cultivated.

The people of the rural areas are used with the cultivation of this culture, and a good tradition has been heritage from one generation to the other.

The olive culture is a big national wealth. A large investment in this sector is a heritage from the past and has been enlarged over the time.

The demand for olive oil and table olives in the domestic market is very high. From the other side, with an adequate technological improvement in the olive processing industry, this product could be traded in the international market.

In the olive production industry, the work is concentrated in the first and the last three months of the year. During this period other cultures need little labour. This will help to decrease the work seasonality in the agricultural sector.

In Albania, according to the Land Agrarian Reform, the olive distribution was done in proportion of the number of people per family and the total number of olives that the former cooperative had. Even in this case (as we saw in the land privatization), we also find a huge fragmentation of the olive plantations. This phenomenon had a big influence in the low olive production and in the quality of manufactured oil.

Olive tree in Albania is cultivated in 22 out of 36 cities of the country and specifically in the regions along the western costal lowland. Geographically we find 3,3% in the plain zone and 96,7% in the hilly zone. Positive is the fact that in 77% of the farms this culture is cultivated in organized plantations whereas in the remaining 23% of the farms this culture is found in a not organized form. The olive concentration in plantations gives the possibility for more careful services and the use of adequate technologies (Kapaj, 2010).

Table 3.1 Olive productions through the years

	2000	2005	2008	2010	2012
Total (1000 heads)	3611	4264	5011	6255	7443
In production (1000 heads)	3256	3488	4179	4298	4576
Yield (kg/head)	11,1	8,6	15,8	16,3	14,3
Production (1000 tons)	36,2	30,2	56,2	70	65,4

Source: Ministry of Agriculture, 2013

As many other sectors of the country's economy, this sector was characterized by a visible depreciation in the main indicators. Huge olive blocks like those in Fier, Mallakaster, Berat, Lushnje, etc. were burned and destroyed. The transformation of the State Farms into private economies in this sector of the economy has been very slow. Even today, there are regions where the reform changes have not yet been completed.

According to the data from Food and Agriculture Ministry, the largest concentration of olives is found in 6 cities of the country; Vlora, Berat, Sarande, Fier, Tirane,

Elbasan. These 6 regions that consist only of 27% of the total number of the cities that cultivate olives, have nearly 70% of the total number of the olive trees and gives 70% of the total olive production of the country. The average yield of production per olive tree in these cities is equal to the average yield per tree in the country level. In the Table 3.2 are given some data about olive production in the main regions of the country and in the Table 3.3 the overall county' situation is described.

Table 3.2: Olive production data for the main regions (2012)

Nr	Region	Number of olives (000 trees)		Yield (Kg/tree)	Production (Ton)
		Total	In production		
1	Berat	628	492	22,0	10841
2	Vlore	532	495	13,0	6436
3	Elbasan	364	331	10,0	3315
4	Fier	347	311	12,7	3955
5	Tirane	318	294	9,1	2664
6	Sarande	312	310	6,6	2048
Σ	TOTAL	2501	2233	13,1	29259
	REPUBLIC	3564	3200	13,1	42012

Source: Ministry of Agriculture, 2013

Table 3.3: Number of heads, yields and olive production according to the regions, 2012

Region	Number of olives (000 trees)		Yield (Kg/tree)	Production (Ton)
	Total	In production		
Berat	628	492	22,0	10841
Delvine	127	126	3,3	419
Durres	57	52	19,3	1004
Elbasan	364	331	10,0	3315
Fier	347	311	12,7	3955
Gramsh	2	2	20,3	31
Gjirokaster	5	4	34,4	150
Kavaje	75	75	13,4	998
Kruje	104	87	4,5	393
Kucove	39	37	20,4	753
Lac	10	10	13,0	126
Lezhe	18	15	9,9	148
Lushnje	227	209	19,5	4070
Mallakaster	197	161	20,9	3362
Peqin	65	64	6,5	412
Permet	2	1	12,8	15
Sarande	312	310	6,6	2048
Skrapar	1	1	11,6	14
Shkoder	93	81	6,0	485
Tepelene	43	43	8,6	373
Tirane	318	294	9,1	2664
Vlore	532	495	13,0	6436
TOTAL	**3564**	**3200**	**13,1**	**42012**

Source: Ministry of Agriculture, 2013

Although there has been a considerable investment in the new olive plantations, the production investments and the services for this culture have been minimal. Today the olive production has low and fluctuating yields. The extensive character of the olive cultivation and the insufficient treatments that are usually done to the olives are the cause of this phenomenon. The yield fluctuation in the olive production has been and still is a serious phenomenon for our country. According to statistical data, the ratio between an "empty" year (year with very low production) and the year with a good production is very high. The Table 3.4 below shows the development of the main indicators in the olive production during 1950 and 2012.

Table 3.4: Number of olives, production and yield, 1950-2012

Year	Number of olive-trees (million)	In production (million)	Production (000 ton)	Yield (kg/olive-tree)
1950	1,7	1,2	6,7	5,5
1960	2,2	1,3	5,0	3,6
1990	5,8	3,4	10	2,9
2000	3,6	3,2	42,0	13,1
2005	4,3	3,4	30,2	8,6
2008	5,0	4,1	56,2	15,8
2010	6,2	4,2	70	16,3
2012	7,4	4,6	65,4	14,3

Source: Ministry of Agriculture, 2013

The olive culture has not been valued as it should. This situation for sure does reflect on the manufacturing industry of this culture, in the quantity and especially on the quality of the olive oil. The olive oil does not meet the quality standards of the "virgin" and "extra virgin" oils. This comes as a result of the non-timely extraction of the oil, olive decomposition before its processing; its processing together with the cores, and above all because of the damages caused by the olive fly. Because of the tradition of the olive oil production and market demands, a lot of important investments and technology improvements are done lately.

3.1.2 Olive age and cultivars in Albania

According to the age of the olives there is a visible distinction that divides the olive plantations into two groups:

Centennial olive plantations are mainly found in the urban areas of Sarande, Vlora, Berat, and Elbasan. These are native varieties with high economic values that consist of the main part of olive production of the country.

Olive plantations planted after the 1960s, which are found by the sea and in the central part of the country.

Based on the statistical data the proportion of the olives according to their age can be done as follows:

Olive plantations above 100 years old (30% of the total olive trees),

Olive plantations from 30-40 years old (45%)

Olive plantations from 10-20 years old (25%).

In Albania nearly 19 olive cultivars are found, but 6 of those have the largest distribution in the country. Table 3.5 below presents the respective percentages of cultivars:

Table 3.5: Olive cultivars

Nr.	Cultivar	Percentage (%)
1	Kalinjot	42,1
2	Large-grained of Berat	17,0
3	Frantoio	7,1
4	Large-grained of Elbasan	6,7
5	Mixan	4,1
6	White olive of Tirana	3,0
7	Others	20,0

Source: Ministry of Agriculture, 2013

3.1.3. Olive harvesting and collecting

Olive collection in Albania starts at the beginning of October and goes on until February. The harvesting is mostly done manually, and no modern equipment is used. During harvesting no selection between olives is done; olives that fall from the wind or as the effect of diseases and olives that are taken from the trees. This way of harvesting has a big influence on the manufactured oil quality.

Transportation to the manufacturing plants is usually done with boxes or big bags. Here, two ways of proceeding are found:

The plant buys the olives and processes it for itself.

The manufacturing plant does the processing against some fee, paid by the farmer and does not buy olive.

In the first case olives are bought by the processing plant according to the quantity. The buying price differs depending on the harvesting period; it increases from October to February. The oil percentage in the fruit, the acidity and other features of the crop are not being considered. And this does not help in improving the oil quality.

In the second case the manufacturing plant takes the service tariff and the oil goes to the bringer.

3.2 Olive oil processing

3.2.1 Olive structure and olive oil production technology

An olive consists of three basic parts: the skin (epicarp), the pulp (mesocarp), and the pit (endocarp). It is made up of about 70 % juice (water and oil; 10 - 60% water and 10 - 30% oil), and about 30% solids on a dry weight basis. The solids are made up of 12 - 25% pit solids, 1 - 3% seed, 8 -10% skin and pulp solids, 3% sugars, 2% proteins, and 2% other compounds such as acids, vitamins, minerals, and pectin. On a dry weight basis, the skin, which represents only about 3% of the fruit weight, contains about 3% oil. The pit represents about 23% of the olive weight and contains about 1% oil. Most of the oil is in the pulp, which represents about 75% of the weight and contains about 50% oil. Not all of the oil can be extracted from the solids with just a physical process, so the solids usually contain about 6-10% oil depending on the variety, maturity, and efficiency of the extraction. The solids contain between 25-70% moisture depending on the processing system used.

The oil fraction is made up of six primary fatty acids: Oleic (55-83%) and Palmitoleic (0.3-3.5%), which are mono-unsaturated; Palmitic (7.5-20%) and Steric (0.5-5.0), which are saturated; and Linoleic (3.5-21%) and Linolenic (0.9-1.5%), which are poly-unsaturated fatty acids. Olive oil is classified as a monounsaturated fat because of the predominance of Oleic acid. Other fatty acids in olive oil at low concentrations are Myristic, Heptadecanoic, Arachidic, Gadoleic, Behenic, and Lignoceric.

Other oil-soluble or semi-oil-soluble compounds in the oil fraction are the waxes, which primarily come from the skin of the fruit. Levels are quite low in virgin olive oil, but appear in higher concentrations when the fruit skins are worked more intensely as in second and solvent extractions of the pomace. The composition and concentration of sterols in olive oil is used primarily to determine its geniuses or authenticity, so that it is labelled correctly in the marketplace. The aliphatic alcohols, hydrocarbons, squalene and pigments give olive oil some of its flavour and colour. They are there in only trace amounts and some are more important than others. The fatty acids, sterols, methyl-sterols, and some alcohols are non-volatile compounds that do not add to the flavour of olive oil, but very important in authentication of olive varieties. The volatile aromatic hydrocarbons and some alcohols are responsible for much of the ultimate flavour of the oil.

When the cell walls are ruptured and all of these oil, water, protein, mineral components, and complex hydrocarbons are mixed together with air, enzymes, and micro organisms the oil fraction absorbs many of the volatile compounds and takes on their flavour and aroma characteristics. The harmony between the oil and water fractions produces the olive oil elixir unique compared to any other fruit product. It is a short-lived, delicate, and positive union that can be amplified, diminished, or disturbed by changing the quality of the fruit, the way it is handled, manipulation of the paste, extraction process, and finally cleaning and storage of the oil.

The negative influence of oxidizing, rotting, and fermenting solids and water must be removed as quickly as possible once the oil has had sufficient contact with the positive aromatic volatiles. Some smaller droplets of oil remain with their protective lipoprotein covering and stay as a fairly stable emulsion within the paste. Some of the oil forms micro-gels; a sort of water-oil-solids mixture colloid, but most of the oil becomes free, rolls around with the water and solids, and begins to unite with other oil droplets.

3.2.2. Technology of olive oil processing

The most common industrial processing method is a continuous extraction system with two centrifugations (first horizontal and then vertical). Vertical centrifugation may be in three phases (as it is shown in the Figure 3.1) obtaining oil, pomace and vegetable waters or in two phases (in this case there is no water injection or little water) obtaining oil an as plastic paste.

Pomace is the solid residual derived after first pressing or centrifugation (a little of olive, pieces of nut, etc.). It may be used for livestock feeding or going through a chemical extraction with the purpose of producing olive-pomace oil.

Vegetable waters are the liquid phase obtained as a result of centrifugation. They are abundant in the three phases extraction method due to the water injection made to the paste before centrifugation. As vegetable waters still contain oil, they are treated a second time in order to get the maximum amount of oil. However, since this is a combination of water and fat, it is difficult to recycle them. Vegetable waters are highly polluting and negatively affect underground water. The most serious ecological problem in olive oil production is the recycling of vegetable waters

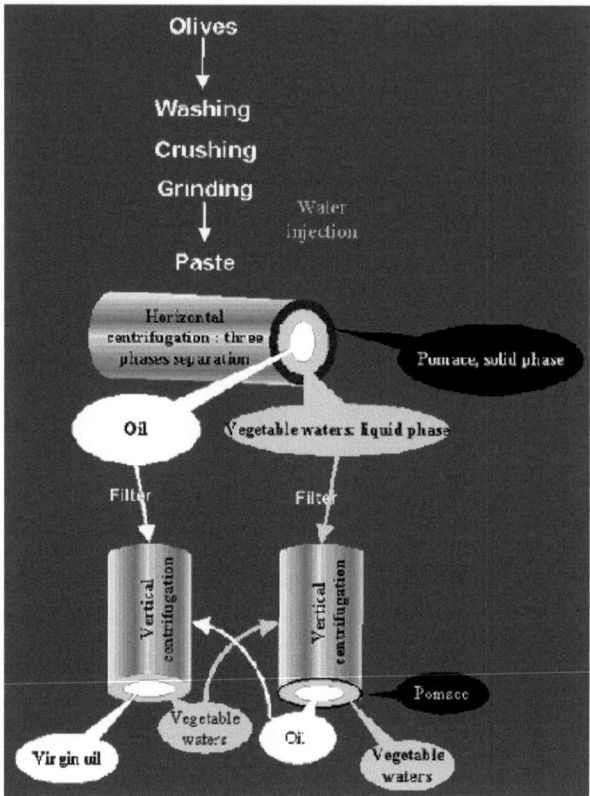

Figure 3.1: Olives processing technology

Source: Vossen, 1999

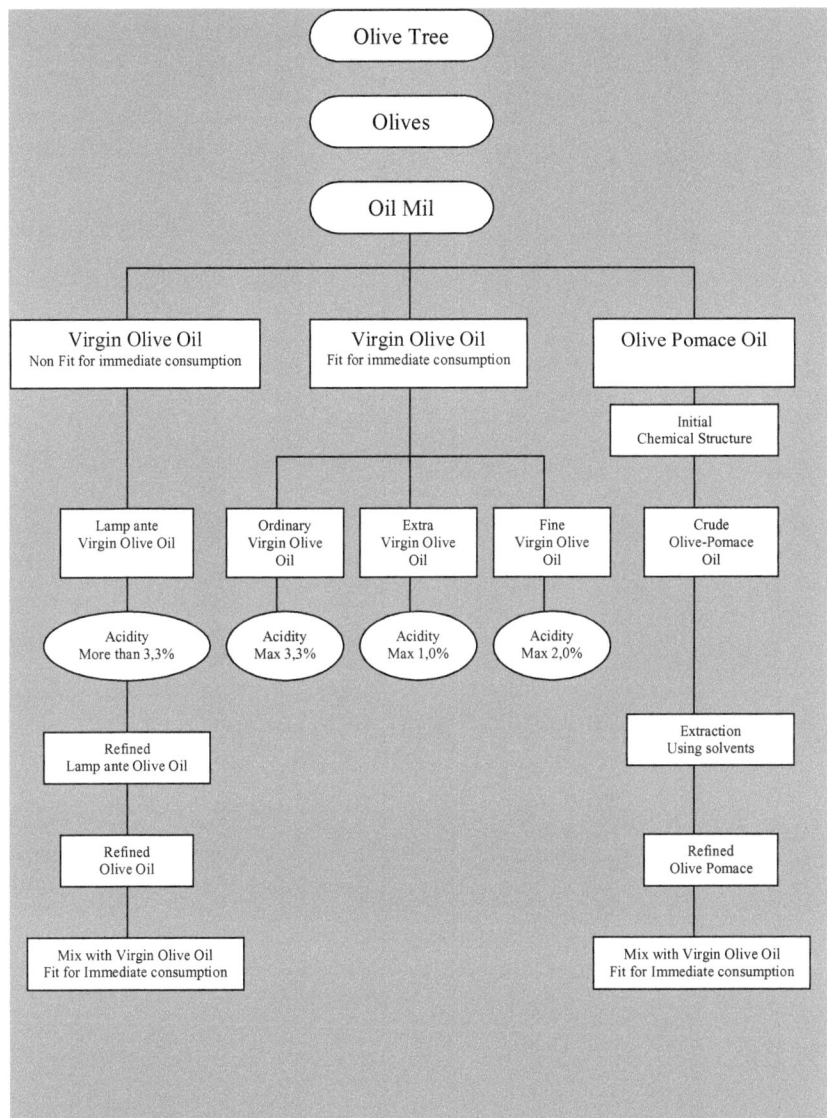

Figure 3.2: Types of olive oil
Source: Adapted from Vossen, 1999.

According to the Figure 3.2 above some definitions in line with the European Standards are given for the specific olive oil type.

1. **Olive oil** is the oil obtained solely from the fruit of the olive tree (Olea europaea L.), to the exclusion of oils obtained using solvents or re-esterification processes and of any mixture with oils of other kinds. It is marketed in accordance with the following designations and definitions:

1.1. **Virgin olive oil** is the oil obtained from the fruit of the olive tree solely by mechanical or other physical means under conditions, particularly thermal conditions that do not lead to alterations in the oil, and which has not undergone any treatment other than washing, decantation, centrifugation and filtration.

   1.1.1 Virgin olive oil fit for consumption as it is includes:

   i. **Extra virgin olive oil**: virgin olive oil which has a free acidity, expressed as oleic acid, of not more than 1 gram per100 grams and the other characteristics which correspond to those fixed for this category in this standard.

   ii. **Virgin olive oil** (the qualifier "fine" may be used at the production and wholesale stage): virgin olive oil which has a free acidity, expressed as oleic acid, of not more than 2 grams per 100 grams and the other characteristics of which correspond to those fixed for this category in this standard.

   iii. **Ordinary virgin olive oil**: virgin olive oil which has a free acidity, expressed as oleic acid, of not more than 3.3 grams per 100 grams and the other characteristics of which correspond to those fixed for this category in this standard.

   1.1.2 Virgin olive oil not fit for consumption as it is, designated lamp ante virgin olive oil, is virgin olive oil which has a free acidity, expressed as oleic acid, of more than 3.3 grams per 100 grams and/or the organoleptic characteristics of which correspond to those fixed for this category in this standard. It is intended for refining or for technical purposes.

1.2. **Refined olive oil** is the olive oil obtained from virgin oils by refining methods which do not lead to alterations in the initial glyceridic structure. It has a free acidity of not more than 0,3 grams per 100 grams.

1.3. **Olive Oil** is the oil consisting of a blend of refined olive oil and virgin olive oil fit for consumption as it is. It has a free acidity of not more than 1 gram per 100 grams.

2. **Olive-pomace oil** is the oil obtained by treating olive pomace with solvents or other physical treatments, to the exclusion of oils obtained by re-esterification processes and of any mixture with oils of other kinds. It is marketed in accordance with the following designations and definitions:

- **Crude olive-pomace oil** is olive-pomace oil intended for refining with a view to its use in food for human consumption, or intended for technical purposes.
- **Refined olive-pomace oil** is the oil obtained from crude olive-pomace oil by refining methods which do not lead to alterations in the initial glyceridic structure. It has a free acidity of not more than 0,3 grams per 100 grams.
- **Olive-pomace oil** is the oil comprising the blend of refined olive-pomace oil and virgin olive oil fit for consumption as it is. It has a free acidity of not more than 1gram per 100 grams. In no case shall this blend be called "olive oil".

3.3. The types of lines and presses used for olive processing

Until the end of the 1970s the olive oil processing was done in traditional primitive ways by the peasants themselves. Gradually with the increases in yield, some plants were built. These were very old technology fashioned plants. Only at the beginning of the 1980s some presses were imported from Italy, and this was the start of innovations in the oil manufacturing plants.

Actually, from 80 manufacturing plants that have been interviewed, almost the half of them uses the "Pieralisi" type presses for the olive oil processing. Figure 3.3 below shows the kind of presses that are mostly used. Second popular kind of press is Alfa Laval with about 15% share of the total. The next significant kinds are Eno Rossi (11%) and Mix (5 %).

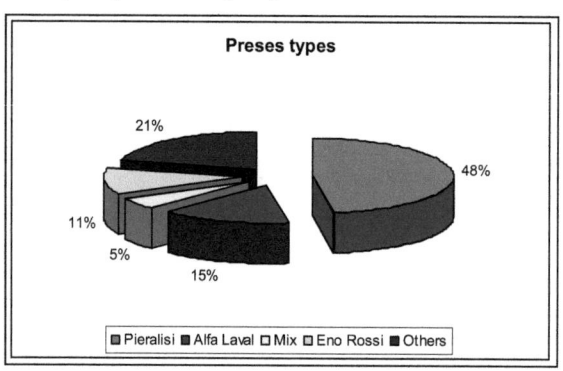

Figure 3.3: Kinds of presses, being utilized in the selected sample plants (%)
Source: Computed based on survey data, 2013

3.4. Type of financing

After the 1990s, a lot of investments were done in the olive oil processing industry. The 80 manufactured plants that were taken into consideration for this study represent almost 75% of the investments done in this sector of the agriculture industry. According to a study done by MaFCP in 2010, the total amount of investments in this sector is 1442,0 million Leke (or 10.686.230,92 euro). In these 80 plants alone, the investments amounted to 1045,0 million Leke (or 7.744.182,6

22

euro). The regions with the highest amount of investment were; Vlora with 25,0% of the total, Tirana with 17, 6%, Saranda with 17,5%, and Fier with 13,8% of the total investments.

All figures of the investments distributions according to the regions are shown in the Table 3.6 below:

Table 3.6: Investments according to the regions

Cities	Million Leke	%
Fier	144,2	13,8
Sarande	182,0	17,5
Vlore	260,8	25,0
Peqin	47,5	4,5
Berat	50,6	4,8
Elbasan	63,7	6,0
Durres	13,3	1,3
Mallakaster	98,3	9,5
Tirane	184,6	17,6
Investment in total	**1045,0**	**100**

Source: Computed, based survey data, 2013

There are three main investment sources in Albania, as far as the agricultural sector is concerned:

Own financial sources

Bank credits

Other funds

The investments are mainly done by the private financial sources of the entrepreneurs. This is followed by a smaller part of those that have taken some bank credits. Figure 3.4 below, shows schematically the share that each of these forms holds in the total investment structure:

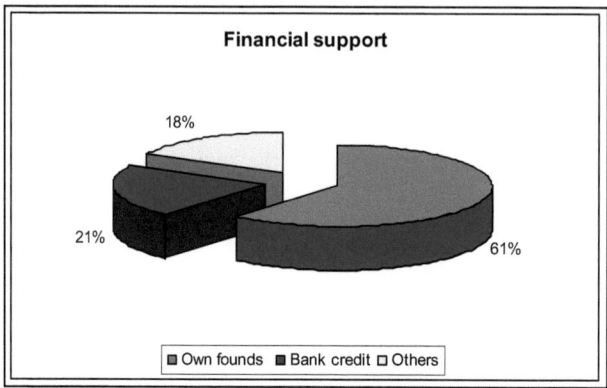

Figure 3.4: Distribution of financial support

Source: Computed, based on survey data, 2013

3.5. Problems related to the marketing of olive oil

Storage

Storage of the product is very important for maintaining the qualities of the oil. This depends a lot on the type of the storage. The accepted forms of storage that are widely used by the producers are inox deposits or plastic boxes. The best way are the inox deposits, which allow the oil quality to be maintained for a longer time. But these are most expensive ones, and this does influence the choice of storage the type.

Distribution

Olive oil distribution is done in two main forms;

- the clients come to the plant to buy the oil,
- oil is distributed to the clients or market by the plant.

The distribution of the product is sporadic and not well organized. Very few of the producers that sell regularly in the market have fixed clients. The others sell the

24

product casually in the market by themselves. Brokers are not commonly used. Only 6 (7,5%) of the surveyed producers do use brokers.

Also the use of contracts is not common among the olive oil producers. From the survey it was seen that only 7 (9%) of the producers make contracts with their clients.

Publicity

For the service business (olive processing), publicity papers are mostly used for stimulating the farmers to bring their olives for processing. Whereas for the olive oil the only means of publicity used until now is the label on the bottle. The survey results discovered that from all the surveyed producers, regularly selling their product on the market, only 7 (9%) use the publicity in their activity.

At the current development stage of this industry the "passing words" through the clients proves to be a very efficient form of publicity. Businesses that have been in this industry for a long time are now superior to those that have entered this industry latter.

Until now no promotional campaign is undertaken by the oil producers.

Packaging and labels

The packaging of the olive oil is done in three forms:

Packaging in plastic boxes of 20-50 litres, is the main form of packaging in this development stage of the processing industry. This is not a good way of maintaining the safety of the product for a long time, because it increases the oil acidity. This type of packaging is mainly used for selling to the wholesalers and to the restaurants.

Packaging in plastic boxes of 5 litres, is mainly used for selling to restaurants and hotels, and sometimes even for home consumption.

Packaging in plastic or glass bottles of 0,5 to 1 litre, is the best way of packaging for protecting the product, but is rather costly and, therefore not so much practised.

According to the data from the survey only 15 (or 19% of the total number) production industries trade their product in a packed form. From these 12 (80%) use plastic and only 3 (20%) use glass bottles.

The label is the best way of publicity used in this industry. It should contain information about the level of the acidity of the oil, the date of production and expiry, olive variety used for the oil, region where the olives comes from, quality and name of the product, and the analysis done to insure the oil quality. The survey found that 10 (12,5%) of the interviewers do use the label on their products. From those, all do put the quality of the oil and their name on the label, but only 8 (80%) of them write the production and expiry date.

Name of the product

The name of the product is very important for the producer because he/she can be identified by the consumers. All those producers, who are using the label, put their name on it as well. It should be mentioned that even though they use names for their products, not all of these are registered. There is the possibility for registering the products in Albania. This will protect the producers from possible falsification of their products.

PART II

4. Literature review, methodological background and analytical results

4.1. Evolution of the concept of comparative advantage

The fundamental reason for international trade taking place between two nations is that both nations benefit from the transaction. If prices are good guides to the alternative use of factors of production, a nation will benefit from international trade when it byes a good from abroad more cheaply than it can buy the same good from domestic sources. The benefit, which might be thought of as an increase in real GDP compared with a situation of no trade, is referred to as the "gains from trade". (Gray, 1987) According to Krugman and Obstfeld (2003), there are two basic reasons why countries trade, leading to these gains. First, countries trade because they are different from each other and secondly, they trade to achieve economies of scale in production.

This part of the study will give a review of theories and concepts of the comparative advantage, which form the methodological background of this research. The Ricardian concept of Comparative Advantage serves as a conceptual basis for a very large part of classical, neo-classical and modern theory of international trade. The theory of international trade is divided traditionally into two disciplines differing in their theoretical framework of analysis:

1. the monetary theory, responsible for the introduction of the concept of absolute advantage, and

2. the pure trade theory, standing for the concept of comparative advantage (Khachatryan, 2000).

Suppose two countries (as described in Ethier, 1988), say Italy and France can produce only two goods, wine and clothes. Each unit of wine produced in Italy requires 8 unit of labour for its manufacture. The Table 4.1 below shows the amount

27

of labour necessary to produce one unit of each good for the two countries. From the data of the table it is evident that both goods require more labour for their manufacture in Italy than in France.

Table 4.1: A simple Ricardian model: Labour required in each country to produce one unit of each   good

Wine	Clothes	
8	4	Italy
1	3	France

Source: Adapted from Ethier, 1988

This makes France having an absolute advantage over Italy in producing both goods, wine and clothes (Ethier, 1988). In France the same unit of product is manufactured with less unit of labour.

Initially each country is producing some of both goods. Then Italy cuts back her production of wine and use the labour released from the wine industry to produce more clothes. According to the example, two units of clothes can be produced with the labour released from the wine production in Italy. On the other hand France stops the clothes production, and the released workforce can be used for the wine production industry. In France exactly three additional units of wine can be produced by the labour transfer.

Table 4.2: A simple Ricardian model: Possible results in each country and in the World

Additional wine	Additional clothes	
-1	+2	In Italy
+3	-1	In France
+2	+1	**In World**

Source: Adapted from Ethier, 1988

The Table 4.2 shows the potential gains that might be attained in each country and in the World. In this way both countries are better of when specializing in the production of one good.

The ratios are then as follows:

a_w^E / a_c^E (8 / 4) and a_w^P / a_c^P (1 / 3), where:

a_w : the amount of labour necessary to produce one unit of wine (in Italy or France)

a_c : the amount of labour necessary to produce one unit of clothes (in Italy or France)

The results depend only on the ratios of the labour requirements within each country and the comparison between them:

$$a_w^E / a_c^E > a_w^P / a_c^P \qquad (8/4 > 1/3) \qquad\qquad (1.1)$$

Thus each ratio tells us the cost, in terms of clothes, of producing one more unit of wine in each country. The inequality says then, that the world must sacrifice more clothes to make wine in Italy than to make it in France. So the conclusion is that as much of the world's wine as possible should be produced in France.

The Equation (1.1) can equivalently be written as:

$$a_c^P / a_w^P > a_c^E / a_w^E \qquad (3/1 > 4/8) \qquad\qquad (1.2)$$

In this form the inequality says that the cost in terms of wine of producing one unit of clothes is grater in France than in Italy. Hence as many of the world's clothes as possible should be produced in Italy. When we speak of the cost of producing wine in the ratio terms of clothes, we are referring to what are termed "opportunity costs". That is, wine is obviously not produced by using up clothes but rather the true cost of wine to the economy is the unit of clothes that can be produced by the labour that is actually used to make the wine, if that labour was instead used by the clothes industry.

⇨ The opportunity cost of a good to an individual or the society is the amount of some other good that must be foregone in order to obtain one unit of it. (Ethier, 1988)

When the labour requirements are in relation shown in the Equation 1.2 we say that "France has a comparative advantage over Italy in wine production relative to clothes" This statement is equivalent to "Italy has a comparative advantage over France in clothes production relative to wine"

The comparative advantage tells us something about the pattern of production. Its implication for the pattern of international trade is even simpler: if trade takes place, a country should export that good in which it has a comparative advantage and import that good in which it has a comparative disadvantage (Ethier, 1988). In our example, Italy should export clothes and import wine. According to the principle of comparative advantage, not the absolute cost differences but comparative costs differences are important factors in determining the trade flows.

The comparative advantage and the gains from the trade are graphically presented in the figure below. The two countries France and Italy produce both good wine and clothes. Their production possibilities curves (PPCs) are AA' and BB' respectively in Figure 4.1. Two sets of community indifference curves represent the demand side of both countries so that E and F are their equilibrium in the conditions of no trade with $U_{0, \text{France}}$ and $U_{0, \text{Italy}}$ levels of welfare respectively. The relative price between goods wine and clothes in the two countries are measured by slopes of their PPCs respectively. According to the figure it is seen that AA' is steeper than BB', meaning that France is relatively more productive in wine or it has a comparative advantage on this good. On the other hand Italy has a comparative advantage in clothes production.

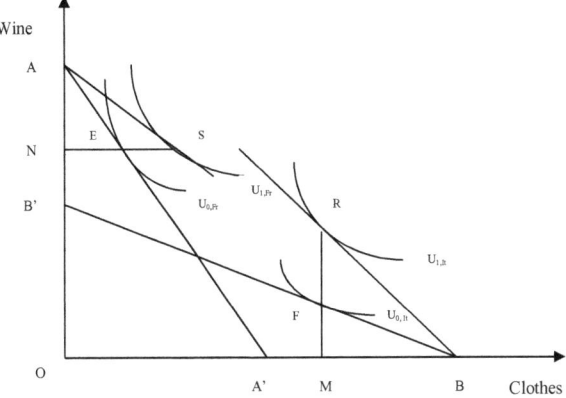

Figure 4.1: Comparative advantage and the gains from trade

Source: Adapted from Kapaj, 2004

Given the difference in their relative price or comparative advantage, the two countries can trade with mutual benefit. France can export wine to Italy and import clothes. Based on the principle of mutual benefit, the exchange price ratio (or free-trade price) must be something in between the price ratios in the two countries. Based on their respective comparative advantage, France specialised completely in wine and Italy in clothes. This means that, in the presence of the international trade, France will produce in point A and Italy will produce in point B. The exchange price ratio between the two countries is presented by the equal slopes of AS or BR as shown in Figure 4.1.

In the presence of the international trade, France which produces at point A consumes at point S and Italy producing at B consumes at point R. The quantity exported by France will exactly equal the quantity imported by Italy and vice-versa. Therefore AN=RM and NS=MB. As shown in Figure 4.1, both countries now enjoy a higher level of welfare at point S and R which are presented by higher indifference curves $U_{1, France}$ and $U_{2, Italy}$ respectively. Therefore both countries have gained from the international trade with each other.

Bhagwati (1998) summarises the features of the Ricardian model as follows:

31

Production: One factor of production whose endowment is fixed. There are two commodities characterised by constant and hence identical marginal and average factor/output ratios in production.

Demand: No specific assumption regarding the demand functions. Trade is balanced, implying that the economy-wide spending equals income.

Trade: There are two counties that can trade the two goods free of transport costs. The factor is immobile internationally.

Market structure: There is perfect competition in all markets.

Supply: In this highly simplified general-equilibrium model, the supply relations are readily derived.

In classical and modern literature the Ricadian simplifying assumptions have been gradually replaced by more realistic ones and thus the theory has become, or so it is hoped, a more adequate model of the real world. (Khachatryan, 2000)

John Stuart Mill (1848) made an extremely important addition to the comparative cost phenomenon by introducing the role of demand and supply forces. He developed the theory of the demand of a country for the products of other countries expressed in terms of units of its exports. In this way he employed the concept of demand elasticity.

Also Marshall has systematically developed the classical theory with the aid of graphical and analytical methods. Marshall's more complex supply and demand curves were the first attempts to represent a general equilibrium in international trade.

The Ricardian theory has been generalized for any number of commodities and countries. Samuelson (1952) has introduced the transportation costs. He suggested a simple way in which transportation costs could be introduced into the two-good models without having to add a third good called "transportation". This is done by adopting the "iceberg" view of transportation: of each unit of a good leaving an origin, only a fraction reaches the destination. The fraction lost in transportation is the transportation cost. This device is used to introduce transportation cost into the Ricardian model.

Judg and Takyama (1971), have shown that the Ricardian principle works for more than two countries, however only when the costs are constant. Whatever the pattern of substitutability and complementarity, the principle of comparative advantage may be reformulated in terms of a correlation between differences in autarky price levels and net export volumes for different commodities (Khachatryan, 2000).

Other prominent researchers, who contributed to the further adjustment of the classical theory of comparative advantage to more realistic situation, are Graham, Lösch and Wilson (1980).

4.2. Further development of the concept of comparative advantage

If labour was the only factor of production, as the Ricardian model assumes, comparative advantage could arise only because of international differences in labour productivity. In the real world however, trade is only partly explained by differences in labour productivity, it also reflects the differences in countries' resources. Thus trade considers not only labour, but also other factors of production such as land, capital and mineral resources. The comparative advantage is influenced by the interaction between nation's resources (the relative abundance of factors of production) and the technology of production (Krogman and Obstfeld, 2003).

That theory advocating, that international trade is largely driven by differences in countries' resources, is one of the most influential theories in international economics and is known as the Heckscher-Ohlin theory. Because the theory emphasises the interplay between the proportions in which different factors of production are available in different countries and the proportions in which they are used in producing different goods, it is also referred to as the factor-proportions theory.

According to the factor-proportions theory, the pattern of comparative advantage is in turn determined by inter-country differences in the relative endowment of primary factors of production. The theory asserts that the comparative advantage of a country lies along the line of products whose productions extensively utilise factors that are relatively abundant in that country, for in a state of autarky their relative prices are expected to be lower than the prices of other primary factors.

The Heckscher-Ohlin theory explains the pattern of comparative advantage in terms of factor endowments as below:

"A county has a comparative advantage in production of the good that uses relatively intensively the country's relatively abundant factor" (Ethier, 1988)

4.3. Introduction to Policy Analysis Matrix (PAM)

The main objective of this study is to evaluate the comparative advantage of the olive oil production in Albania. For the fulfilment of the research objective, the DRC ratio for olive oil production was estimated within the framework of the Policy Analysis Matrix (PAM). The general framework of the PAM is given in Table 4.3 below followed by the specific explanations.

The method distinguishes private and social profitability. Private profitability (shown in the first row of Table 4.3) is determined using actual input and output prices prevailing in the domestic market.

Private value: Private value refers to actual, observed values for revenues and costs. These values reflect the prices received or paid by the farmers or processors in the agricultural system. The private prices incorporate the economic costs or values plus the effects of all distorting policies and the market failures.

Social profitability that is shown in the second row of the PAM table provides a measure for the comparative advantage.

Social value: The social value measures the comparative advantage or the efficiency of a commodity in the agricultural system. These values are an efficient measure because the inputs and outputs involved in the system are valued at their social opportunity costs. When we talk about the private value we also have to deal with the effect of the distorting policies, effect which the social value eliminates.

Table 4.3: Policy Analysis Matrix

Item	Revenues	Cost of Tradable Inputs	Cost of Non-tradable inputs	Profits
Private Prices	A	B	C	D
Social Prices	E	F	G	H
Divergences	I	J	K	L

Source: Nguyen, 2002

Note, where:

A: Revenues at private prices

B: Cost of tradable inputs in private prices

C: Cost of non-tradable or domestic inputs in private prices

D: Private profits $\{D = A - (B + C)\}$

E: Revenues at social prices

F: Cost of tradable inputs in social prices

G: Cost of non-tradable or domestic inputs in social prices

H: Social profits $\{H = E - (F + G)\}$

I: Output transfers $(I = A - E)$

J: Input transfers $(J = B - F)$

K: Factor transfers $(K = C - G)$

L: Net transfers $(L = D - H$ or $L = I - J - K)$

By looking at both private and social prices, the PAM allows us to determine the extent of profitability and divergences resulted from policy distortions and market failures in both input and output markets. The values of parameters in the last column and last row in PAM Table 4.3 carry these economic meanings which are summarized as follows (Nguyen, 2002):

- Private profit (D): This parameter gives a measure of private profitability which is the difference between the revenue (A) and costs (B+C) measured in private or actual market prices. The sign of D shows the producer's profitability. If D is

35

positive the producer is profitable in producing the product and he is unprofitable when the sign is negative.

- Social or economic profit (H): This is an efficiency measure as it is the profit measured in social or economic terms. All the output E and the inputs F+G are valued at their social opportunity costs which reflect the scarcity of the resources. If H is positive it is desirable for the country or the society to produce the product and vice-versa. It is noted that this parameter needs not to have the same sign as D, as the product might give profit to the private producer but can be socially unprofitable.

- Output transfer (I): This parameter measures the divergence in terms of revenue. This is the difference between the private revenue (A) and the revenue in social prices (E). It shows the level of policy distortion in the product market. If it is positive, it implies that the private revenue is greater than its social counterpart, the producer is supposed to receive a subsidy. On the contrary, if it is negative, it is equivalent to a tax on the producers in their production.

- Tradable input transfer (J): Measures the extent of divergence between the private and social cost values of tradable inputs as a whole. It reflects the policy distortions, causing the divergence between the domestic and world prices of the inputs. If J is positive implying that the private cost of tradable inputs are over their social values, the producers are equivalent to being taxed and vice-versa.

- Domestic factor transfer (K): The meaning of this parameter is similar to the tradable input transfer, except it is applied for the domestic factors as a whole. If this parameter is positive meaning that the private values of domestic factors are greater than their social counterparts, the producers are taxed on their uses of domestic factors. If K is negative, the actual costs of the country in terms of domestic factors are larger than the producer pay. In this case, the producers can be said to have a subsidy on their domestic factor uses.

• Net transfer (L): This parameter measures the total of net distortions in both the input and output markets. This net transfer from distorting policies is the sum of all distortions in the commodity, tradable input, domestic factor and foreign exchange markets. If L is positive it implies that the product is more profitable privately than socially. If it is negative the product is more socially profitable than privately.

PAM can be considered as a simple general equilibrium and policy-oriented simulation model. It is just a simple general equilibrium model due to its static nature. The greatest advantage of PAM is that it allows the disaggregation of the production activities and their costs. The cost components are examined directly and to a very detailed degree. In this sense the PAM is quite close to a partial equilibrium model because it moves down to a very detailed level within an economic sector. In the PAM the indicators of policy distortions and economic efficiency are estimated on a straightforward basis resulting in relatively reliable outcomes. The PAM also allows testing of a wide range of policy options for production systems.

The PAM however inherits some limitations and it does not examine explicitly the economic relations between sectors of the economy and dynamic effects of the policy. The PAM assumes fixed levels of macroeconomic variables as well. In order to minimise the weakness of the PAM, sensitivity analysis is done in this study allowing for variations in the world input and output markets, macroeconomic variables such as exchange rate and agricultural policies.

From the PAM results there are some indicators of distortions and comparative advantage that can be derived. In this study only two of these indicators are evaluated - Private Cost Ratio (PCR) and Domestic Resource Cost (DRC). The estimation of these two indicators will help us to see if the olive oil production in

Albania is privately and socially profitable. More emphasise is given to the DRC ratio as it is used to measure the comparative advantage of the olive oil production.

4.4. Study area

Albania has 126 olive oil processing plants (MaFCP, 2010). These are mostly spread in the central and south-western regions of the country and very little are found in the north part by the Adriatic Sea.

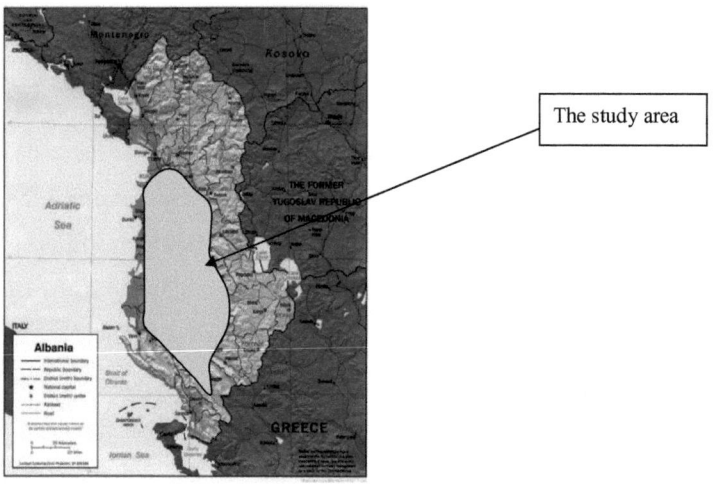

Figure 4.2: Map of Albania showing the study area

Source: www.shwiperia.com, 2014

For carrying out this study information from 80 processing plants was collected. They are all positioned in the central and south-western part and the olive oil production taken from these plants represent more than 80% of the total olive oil production of the production year 2012-2013.

4.5. Private Cost Ratio

The private cost ratio (PCR) explains the ratio of domestic factor cost (C) to the value added in private prices (A-B). This ratio demonstrates the ability of the production system to cover the cost of the domestic factors and continue to be competitive. It is also a proxy for the degree of processing within the domestic economy. This ratio is important for investors because they can optimise their profits by minimising the costs of tradable inputs and factors.

PCR = Cost of Non-Tradable Inputs / (Revenues – Cost of Tradable Inputs)

= C / (A - B)

From the calculation of the collected data, the following results were observed:

PCR = C / (A – B) = 13183, 69 / (80271, 3 – 59288, 07)

PCR = 0, 63

The result indicates that olive oil production in Albania is privately profitable because the ratio is between the limit intervals {0-1}.

4.6. Conceptual framework of the DRC analysis

To estimate the comparative advantage of a commodity, in this case the production of the olive oil, this study implements the methodology of the DRC estimation described by Monke and Pearson (1989), as a ratio of the opportunity cost of domestic factors of production per unit of value added in world prices.

The DRC ratio is calculated using the formula:

$$DRC = G / (E - F) = \Sigma\, W_pF_p / (\Sigma\, P_cF_c - \Sigma\, P_iT_i)$$

Where:

W_p: social price (opportunity cost) for domestic resource or non-tradable inputs,

F_p: coefficients of domestic resource,

P_c: border price for tradable outputs,

F_c: quantity produced of tradable outputs,

P_i: border price for tradable inputs,

T_i: coefficients for tradable inputs.

4.7. Steps of calculating the DRC

Calculating the DRC of a commodity involves 5 main steps:

-   activity selection
-   budget construction
-   inputs and outputs classification
-   social price calculation
-   sensitivity analysis conduction

A short description of these steps is presented in the next paragraphs.

Activity selection

For the moment Albania is still in a phase where markets are in a process of transition and far from being competitive. So the actual market prices are more likely to represent market failures rather than marginal costs. A calculation with only the actual values of olives and olive oil production will therefore be misleading. Instead an estimation of the competitiveness or the comparative advantage of these commodities, with economic or social values is used. This will reflect a situation less affected by the market failures and the distorting policies.

The study focused on the olive oil production. Olive oil, as a final product, and the production input (olives), are both commodities available in the world market. There are enough investigations stating the existence of comparative advantage in the olive production in Albania (Mance, 2002). Thus the olive production can be omitted and olives considered as a tradable input for the processing activity.

Budget construction

The budget construction includes the calculation of all activities of olive oil processing in Albania, which is actually the estimation of the olive oil production costs at an olive oil manufacturing plant. Here all the expenses and the returns of the plant were considered. The budget calculation is done based on the information gained from the specialists of the oil processing plants and are verified with the existing budgets of the previous years.

Input and output classification

Classifying the inputs and the outputs into tradable and non-tradable is the next step in the DRC analysis.

> ➤   Tradable commodities are those inputs and outputs of production, which can be traded internationally:
>
>   E.g., olives and olive oil, fuel, machinery, packaging materials (bottles, labels and boxes) are tradable inputs. Even though some of these products are are produced in Albania they are considered as tradable as far as they enter the international market.

> ➤   Non tradable commodities are those production inputs which cannot be traded internationally:
>
>   E.g., land, labour, water and capital are factors that do not enter the international market, and are considered as non-tradable.

Some cost items, however, contain both tradable and non-tradable components and in some cases transfers (taxes, subsidies) as well. Such costs are further disaggregated and classified into tradable, non-tradable and transfer components.

Social price estimation

After construction of the production budget using the actual market prices (private prices) of the commodities and after classifying all the budget entries into tradable and non-tradable, the next step is to estimate the social prices. Social prices are intended to reflect the true economic value of inputs and outputs in the absence of taxes, subsidies, tariffs and quotas, price control and other effects of government policies, market imperfections and externalities.

The social prices are expressed in European Currency "Euro (€)" using the official exchange rate of the study period 202-2013.

Social price of tradable inputs and outputs

The social price of an agricultural commodity is the border price, the price at which foreign suppliers would deliver the commodity to the domestic market or the price that foreign consumers would pay domestic suppliers to deliver the commodity to their markets (Monke and Pearson 1989). The social value of additional domestic inputs is thus the foreign exchange saved by reducing imports or earned by expanding exports.

In this study the Italian market is assumed as reference point where Albania imports fuel, bottles, boxes, labels and other tradable inputs. The appropriate social values of tradable inputs and outputs are given by world prices; C.I.F. (cost, insurance and fright) import prices and F.O.B. (free on board) export price for tradable goods and services.

Social price of non-tradable inputs

The social value of each of the non-tradable factors is found by estimating the net income forgone because the factor is not employed in its best alternative use. What does the economy forgo because input X is used in the production of Y? There are two polar cases. First, no alternative is forgone; if X is not used in production of Y, it stays idle; there is surplus of X relative to available opportunities. Second, an alternative is forgone; and the contribution of X in the alternative output Z would have been as valuable as in the current production of Y. The task is to identify and price this best alternative. (Khachatryan, 2002)

Labour

Despite the rather high rate of urban unemployment and differences in wage levels between regions and sectors there are no interregional labour movements in Albania, because of high costs of travelling and housing. So the labour is considered as a fixed factor.

For evaluating the social price of labour the opportunity price of labour, i.e., the next best alternative employment was assessed. The best alternative was the actual wage rate in meat processing industry and this was taken as the social value for labour.

Capital

For the evaluation of the social price of the capital, the average annual interest rate was used. The best alternative for the olive oil producer was considered to put the money in the bank and get the annual interest from it.

Water

An estimation of social value of water the water prices in the other sectors of the processing industry were taken.

4.8. Estimating the shadow exchange rate

One of the most important tasks in the DRC methodology is to estimate the shadow exchange rate in order to derive the economic import parity price of imported inputs and the economic export parity price for the output, which in this case is the olive oil. A general and most simple definition of exchange rate is the price of a unit of foreign currency in terms of domestic currency. The foreign currencies chosen in economic analysis are internationally traded currencies (in this study the Euro is chosen). The exchange rate that one can see is the market exchange rate or official exchange rate. In a regulated exchange rate regime, there is often a difference between the regulated exchange rate and the market exchange rate.

The official exchange rate is affected not only by market forces but also by economic policies. It means that it contains all the policy distortions especially the trade distortions. To examine the comparative advantage of a country for a product, these distortions should be removed in the official exchange rate. That is why the concept of shadow exchange rate is introduced. The shadow exchange rate (SER) can be defined as the rate of exchange which accurately reflects the consumption worth of an extra unit of foreign exchange in terms of domestic currency. (Brent, 1998)

In this study the shadow exchange rate, was estimated by adapting the approach used by Nguyen (2002). Here the free trade exchange rate (FTER) was used as the shadow exchange rate. As the impact of trade liberalisation is considered, the FTER is the most appropriate rate in the analysis. From here, SER and FTER were used interchangeably.

The free trade exchange is the exchange rate which would prevail if all restrictions on trade are removed, its formula according to Brent (1998) is:

$$SER = OER / SCF \qquad (1.1)$$

Where:

SER is the shadow exchange rate (or FTER)

OER is the official exchange rate, and

SCF is the social conversion factor.

Following the formula transformation, the social conversion factor is:

$$SCF = OER / SER = \{Ex + Imp\} / \{(Ex - t_{ex}) + (Imp + t_{im})\} \qquad (1.2)$$

Where:

Ex are export values in F.O.B. prices

Imp is import values in C.I.F. prices

t_{ex} and t_{imp} are average export and import tax respectively.

Equation 1.2 is used for estimating the shadow exchange rate for Albania in this study. This equation suggests that if average import and export taxes are equal to zero, the SCF = 1 and this will lead to the conclusion that SER = OER implying that there is no distortion in the foreign exchange market. If the average import tax increases, the SCF will be lower, resulting in a higher figure for the shadow exchange rate and vice-versa. By contrast if the average export tax increases, the corresponding SCF will also increase resulting in a lower SER.

The value for "Ex" used in the formula is the average of total export values in the year 2012 and 2013 and the "Imp" value is the average of total imports in these two years. The value for the import taxes were taken from the Ministry of Finance. For the export taxes, according to the Albanian law, no export taxes, export bans or measures with equivalent effect are being used. Having found all the above values, Albanian's SCF during the period 2012-2013 can be measured as followed:

$$SCF = \{Ex + Imp\} / \{(Ex - t_{ex}) + (Imp + t_{im})\} = 0, 9$$

Once measuring the social conversion factor (SCF = 0, 9), the other step is to evaluate the shadow exchange rate. This calculation considers the official exchange rate of the period October 2012 – March 2013 (period in which the olive oil production is done in Albania). The shadow exchange rate was found:

$$SER = OER / SCF = 134,94 / 0,9 = 149,93 \text{ Lek} / €$$

44

4.9. Estimating the import parity prices of tradable inputs and the export parity price for olive oil

4.9.1. Import parity price for tradable inputs

The most important tradable input in olive oil production is olive. For olive oil production in Albania, mainly used are olives produced in the country, but for evaluating the social price of olives, it was supposed that the olives are imported from Italy. So the import parity price of olives at the plant gate has been calculated. For the calculations was started from the F.O.B. price at export point, in Italy. Than with adding all the costs for freight, insurance and unloading at import point, in Albania, the C.I.F. price at point of import was found.

The C.I.F. olive price was then converted into the local currency using the official exchange rate (OER). Adding all other expenses like costume taxes, port charges, transport, and traders' profit, multiplied by the standard conversion factor, the social price of olives at the plant gate is calculated.

All the estimation results are shown in the Table 4.4 below:

Table 4.4: Calculating the import parity price for olives

(take)	F.O.B. price in Italy (in €)	20,00
(add)	Fright, insurance and unloading	0,30
(equal)	C.I.F. price in Albania (in €)	20,30
(convert)	Convert to local currency with OER	2.739,28
(add)	Local port charges x SCF	0,72
(add)	Local transport to the trade x SCF	0,54
(add)	Wholesaler profit x SCF	123,30
(add)	Retailers profit x SCF	257,75
(add)	Transport to the plant x SCF	0,72
(equal)	Import parity price at the plant	3.122,31

Source: Computed data 2012-2013, table adapted from Nguyen 2002

4.9.2. Export parity price of olive oil

For estimating the export parity price of olive oil it was supposed that Albania exports olive oil to Italy. The derivation of the export parity price for oil at the plant gate started with the C.I.F. price at import point in Italy. All the costs involved for fright, insurance and unloading were deducted to get the F.O.B. price at point of export in Albania. The F.O.B price of olive oil at the export point is then converted to local currency using the official exchange rate (OER).

Then all the costs involved from the port to the plant gate multiplied by the standard conversion factor were deducted to get the social price of olive oil at plant gate. All the calculations for estimating the export parity price of olive oil are shown in the Table 4.5 below:

Table 4.5: Calculating the export parity price for olive oil

(take)	C.I.F. price in Italy (in €)	2,10
(deduct)	Fright, insurance and unloading	0,30
(equal)	F.O.B. price in Albania (in €)	1,80
(convert)	Convert to local currency with OER	242,89
(deduct)	Local port charges x SCF	0,72
(deduct)	Local transport to the trade x SCF	0,54
(deduct)	Wholesaler profit x SCF	10,90
(deduct)	Retailers profit x SCF	20,77
(deduct)	Transport to the plant x SCF	0,72
(equal)	Export parity price at the plant	209,25

Source: Computed data 2012-2013, table adapted from Nguyen 2002

4.10. DRC ratios and sensitivity analysis for olive oil production

4.10.1. The interpretation of DRC results

The DRC results indicate whether the production of a commodity has a comparative advantage for a given country. It does reveal the efficiency of the use of domestic resources to save (or earn) one unit of foreign exchange. The interpretation of possible results is given in Table 4.6 below:

Table 4.6: Interpretation of DRC ratios

DRC ratios	Interpretation	Conclusion
DRC = 1	The economy neither gains nor saves foreign exchange through domestic production	Economy in balance
0 < DRC < 1	Value of domestic resources used in production is less than value of the foreign exchange earned or saved	Comparative advantage
DRC > 1	Value of domestic resources used in production is grater than value of foreign exchange earned or saved	No comparative advantage
DRC < 0	More foreign exchange is used in production of a commodity than the commodity is worth	No comparative advantage

Source: Khachatryan, 2002.

From the calculations of the data for the DRC ratios the following results were observed:

$$DRC = G / (E - F) = \Sigma \, W_p F_p / (\Sigma \, P_c F_c - \Sigma \, P_i T_i)$$

$$= 14.237,210 / (56.937,71 - 39.862,48)$$

DRC = 0,83

The DRC ratio equal to 0,83 shows that Albania has comparative advantage in the olive oil production even though the efficiency is not very high. The estimated DRC ratio indicates that the value of domestic resources used in the production of olive oil is less than the value of foreign currency earned or saved.

4.10.2 Sensitivity analysis

Sensitivity analysis is a good tool for revealing the changes in the comparative advantage rankings, when the individual parameters change. It can be used to assess the effects of possible errors in evaluation of technical coefficients of the production plant budgets, or errors in the social prices. This analysis is done to examine the effects of the changes in parameters like yield and production, world reference prices of commodities, wages and exchange rates. The DRC ratios have been calculated changing the values of the basic model parameters at 20% into both directions to asses the impact of possible changes.

The sensitivity analysis has been carried out to identify how the DRC ratios for olive oil production react to various parameter changes. The robustness of the DRC has been checked against the variation of the following parameters: prices in the international markets used for calculation of social prices of tradable, labour wage and exchange rates. The results of calculations of DRC ratios for olive oil production involving three different scenarios, which reflect possible changes of model parameters, are presented in the Table 4.7 below.

Table 4.7: Results of sensitivity analysis in DRC ratios

SCENARIOS		DRC Ratios	
Parameters	% changes	Scenario	BASE RUN
Price of Olive oil	0, 8 * base	2, 50	0, 83
	1, 2 * base	0, 50	0, 83
Price of Olives	0, 8 * base	0, 57	0, 83
	1, 2 * base	1, 52	0, 83
Labor	0, 8 * base	0, 80	0, 83
	1, 2 * base	0, 87	0, 83
Exchange rate	0, 8 * base	1, 07	0, 83
	1, 2 * base	0, 68	0, 83

Source: Adapted from Khachatryan, 2004 and computed data 2012-2013

The DRC ratios show a high sensitivity towards the world prices. When the world market price of olive oil falls by 20% of the actual one the DRC ratio increases drastically (2,5), making the olive oil production no more competitive. In the case of 20%, higher world prices of olive oil, the olive oil production becomes even more competitive in Albania, attaining a DRC ratio of 0,50. Similarly the world price for olives was changed. When the price decreases with 20% the DRC ratio decreases too (0,57 compared to 0,83 in the base run), making the olive oil production more competitive. The opposite happened when the olive price increases by 20%, the product is becoming more competitive (1,52).

Olive oil production is not considerably influenced by differing cost of the labour force. The market changes for the labour force factor bring up minimal, almost intangible changes in production efficiency. A 20% change in the labour force price in both directions influence the production efficiency with only 3%.

As it is seen from the Table 4.7 olive oil production is also sensitive to the respective changes in the exchange rate. Firstly a change of the official exchange rate (reducing it by 20%) is considered. This change influenced olive oil production negatively. With a DRC ration equalling to 1,04 the production is no more competitive. Whereas when the exchange rate was changed in the other direction, getting 20% higher, then the olive oil production becomes more competitive.

Even though from the DRC calculations prove that olive oil production in Albania has a comparative advantage, is however not very strong. In those scenarios of the sensitivity analysis, where a slight deterioration of the production conditions are predicted, the DRC ratios for olive oil production change easily turning the industry into a non competitive one.

PART III

5. Olive Oil production in the international trade

5.1. The Olive Oil production in the international trade

European countries, particularly those in the Mediterranean basin, dominate world production and consumption of olive oil. This region is home to the olive tree, with olive oil playing a major part in the Mediterranean diet and culture. Post Second World War migrants from these regions have helped spread the tree's distribution and introduced olive oil to the diets of other countries. Now there is widespread interest in producing olives for import substitution and regional trade development.

Olive oil production is concentrated in the Mediterranean basin countries: Spain, Portugal, Italy, Greece, Turkey, Tunisia and Morocco. These seven countries alone account for 81% of world production. Spain, Italy and Greece are the world's leading producers and exporters of olive oil, collectively representing nearly three-quarters of world production ranked respectively in 2013. World olive oil production has increased by an average of 11% per annum since 1995. The main contribution to this increase was in 1996, which followed a year of drought in the Mediterranean. The Figure 5.1 below shows the major olive oil producing countries in the World in 2013.

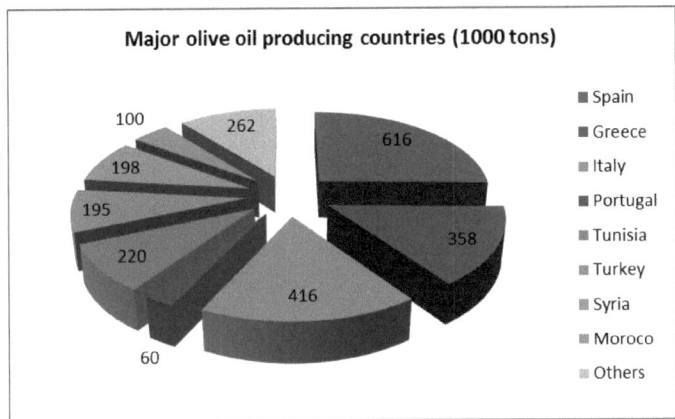

Figure 5.1: Major olive oil producing countries
Source: http://www.internationaloliveoil.org/, 2014

A look at the exports indicates that the main exporting countries are situated in the Mediterranean region. The export flows for the years 2012-2013-2014 are given in the Figure 5.2.

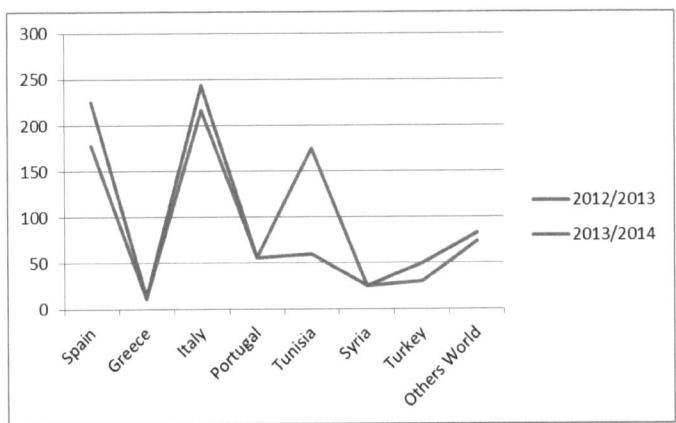

Figure 5.2: Exports of olive oil, 1000 tons
Source: http://www.internationaloliveoil.org/, 2014

In the table 5.1 below are given in more details the export trends in the last three years. We can see that the majority of the exports come from European countries.

Table 5.1: Olive oil exports through the world

World Olive oil exports (1000 tons)		
	2012/2013	**2013/2014**
Spain	177,5	225,00
Greece	11	13,00
Italy	216,40	243,00
Portugal	56,00	55,80
Others EU	1,60	1,70
Total EU	**462,5**	**538,5**
Tunisia	175,00	60,00
Syria	25,00	25,00
Turkey	30,00	50,00
Others World	72,50	81,00
TOTAL WORLD	**765**	**754,5**

Source: http://www.internationaloliveoil.org/, 2014

A look at the imports indicates that the main importing country is US. The import flows for the years 2012-2013-2014 is given in the Figure 5.3.

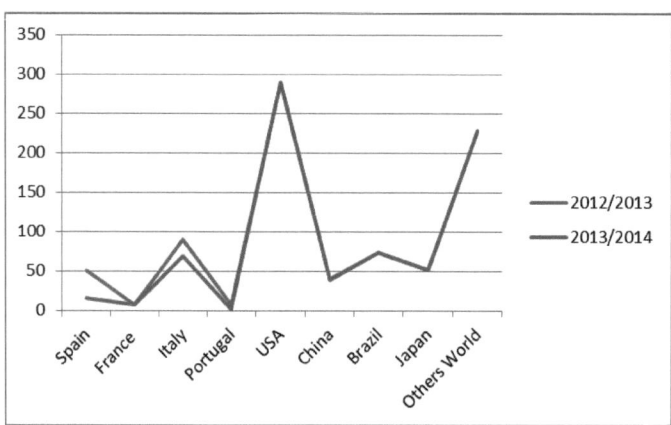

Figure 5.3: Imports of olive oil, 1000 tons
Source: http://www.internationaloliveoil.org/, 2014

In the table 5.2 below are given in more details the import trends in the last three years.

Table 5.2: Olive oil imports through the world

World Olive oil imports (1000 tons)		
	2012/2013	**2013/2014**
Spain	50	15,00
France	6,80	7,20
Italy	90,00	69,00
Portugal	7,00	1,00
Others EU	1,70	0,80
Total EU	**155,5**	**93**
USA	288,00	290,00
China	39,00	40,00
Brazil	73,00	73,00
Japan	51,00	51,00
Others World	226,00	227,00
TOTAL WORLD	**832,5**	**774**

Source: http://www.internationaloliveoil.org/, 2014

The world olive oil production during the last three years (2012-2013-2014) is shown in the Figure 5.3 below. Production trend by country is ascending but the great influence of the three major producing countries introduced a high level of uncertainty in the production level. Indeed, the fact that production in Greece, Italy and Spain changed much more than one of the other producing countries explains the high volatility of global production.

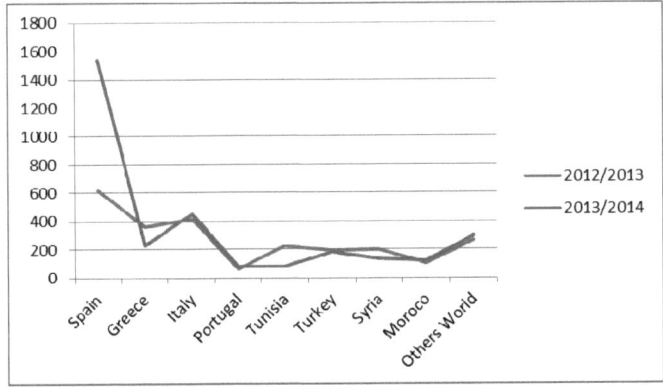

Figure 5.4: Olive oil production, 1000 tons
Source: http://www.internationaloliveoil.org/, 2014

Finally, it should be mentioned that the production of other countries such as Australia and the United States, is increasing. Main consuming countries are also the main olive oil producers, as can be seen from the Figure 5.5. The European Union accounts for 55% of world consumption. Mediterranean basin countries (Italy, Spain and Greece) represent 43% of world consumption. Other consuming countries are the United States, Morocco, Syria and Turkey.

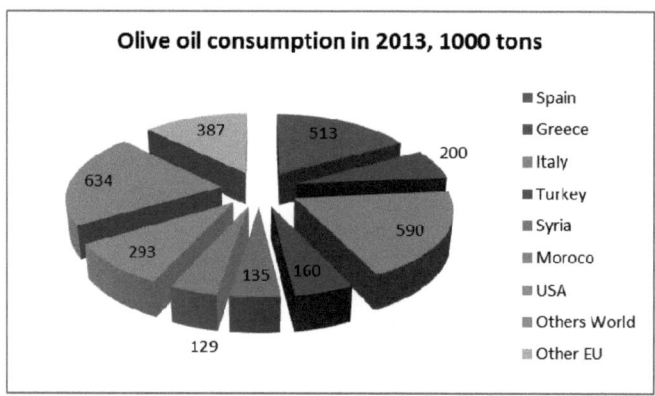

Figure 5.5: Olive oil main consuming markets
Source: http://www.internationaloliveoil.org/, 2014

The evolution of production and consumption shows a slight growth from the 1970s to the early nineties. In the mid 1990s there was a strong increase both in production and consumption. Despite the fall in production that came afterwards, consumption did not decrease.

The strong correlation between European and world consumption trends explains the significance of European consumption. However, the increasing convergence recently observed between the two trends (Production and Consumption) is the result of the emergence of new markets for olive oil. In the Figure 5.6 are given the production and consumption lines in World and Europe.

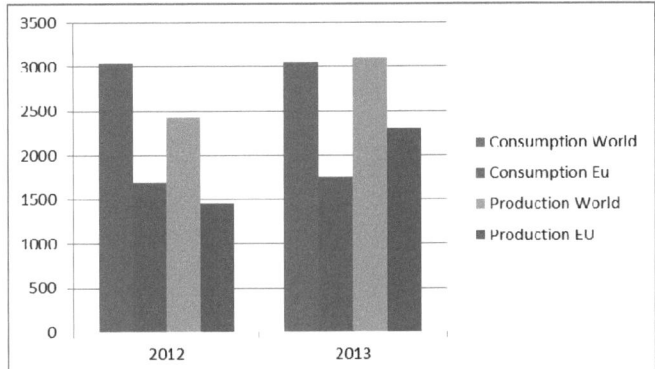

Figure 5.6: Production and consumption of olive oil in the World and in Europe
Source: http://www.internationaloliveoil.org/, 2014

New economic support systems are being established in the European farming sector. An increase in the overall quality of oils, due to EU aid to Spain being dedicated to quality improvement, could have a gradual effect on prices. Whether the increased supply of higher quality oils raises or lowers the average price will depend on changes in consumer preference for oil quality.

The present structure of the production and marketing chain is influenced by the importance of olive oil crops in the Mediterranean area.

1. Production is highly fragmented. It is divided into a number of small properties, which grow olives for oil production. This is particularly the case in Spain, Italy, Greece and Portugal. These farms are organized into cooperatives in order to process the olives and obtain oil or sell the olives directly to processing companies.

2. Refining operations are more concentrated. However, in the case of Spain there were 80 refining companies (cooperatives included).

3. The market is very competitive and entry barriers are quite strong.

5.2. Olive oil situation in three major producing countries

As mentioned earlier, Spain, Italy and Greece are the three leading countries in the olive oil production. The study employs the reference prices and indicators of these countries, as places within the same region and with similar climatic conditions for the olive cultivation. For this reason a short description of the markets of these countries is given below.

5.2.1. Spanish Market:

Spain has 4,7 million acres of olive trees under cultivation, which ranks it as the top producer and exporter of olive oil in the world (49% of olive oil production in 2013) with a production of approximately 1.536.000 tones of olive oil. By comparison, Italy the second ranked producer has 2.0 million acres. Although Spanish olive acreage has been steadily declining since the 1960s, new plantings in recent years have altered the trend, particularly in Andalucia region, which produces approximately 80% of Spain's olive oil overall. Due to the gradual replacement of order low-producing orchards with higher density and more productive orchards, average production per acre has been rising. In Table 5.1 below are given some data on the olive production in Spain in the last years.

Table 5.1: Olive production in Spain, 2000-2013

	2000	2002	2005	2008	2010	2012	2013
Production							
(1000 tons)	669	1411,4	989,8	1236,1	1401,5	1615	616,3

Source: http://www.internationaloliveoil.org/, 2014

Oil production has been in its maximum production in 2012 with 1.615.000 tons. Production is highly influenced by seasonal rainfall (most of the acres are not irrigated), and alternate bearing (low yields followed by higher yields due to the influence of crop on the next year's production). Spain is also a large olive oil consumer (Figure 5.7)

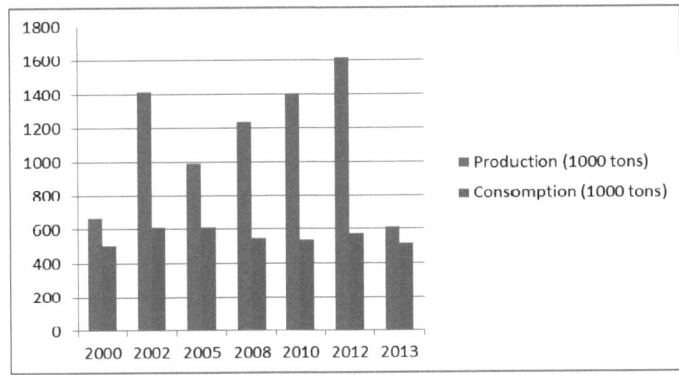

Figure 5.7: Olive oil consumption and production in Spain
Source: http://www.internationaloliveoil.org/, 2014

National average consumption of olive oil is 557.700 tones, based on the years 2000-2013, and has been moderately increasing in Spain. As a comparison Italy consumes more oil that it produces (production = 575.000 tones and consumption = 695.000 tones, average for 2000-2013). Spain exports much of its excess production in bulk to other European Community Countries where it is consumed or repackaged and exported. In the Figure 5.8 below it is given a view of the import-exports of olive oil in Spain from 2000-2013.

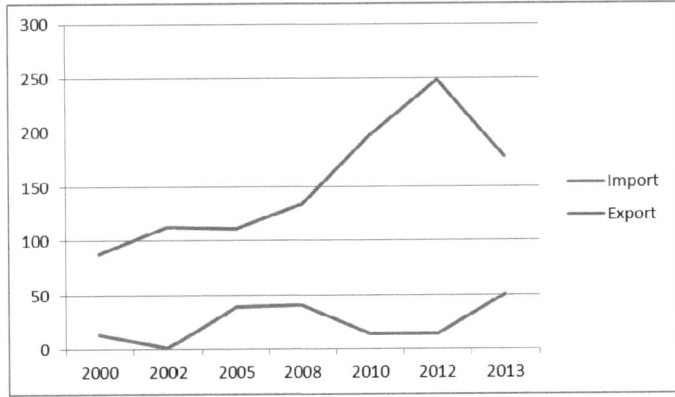

Figure 5.8: Import-export situation of olive oil in Spain, 2000-2013(1000 tons)
Source: http://www.internationaloliveoil.org/, 2014

It is the Autonomous Community that carries the most weight in the Spanish olive market, therefore, where more wastes are produced and also where more energy exploitation plants based on these wastes can be found. Spain, along with the European Union, has a national marketing campaign to boost the generic consumption of olive oil through promotion of the healthful aspects of a Mediterranean diet. Brand advertising is based on personal taste, confidence in companies and price aggressiveness. Price is the predominant factor in Spain, though quality is important.

The Spanish and Greek industries are better positioned due to the aggregated nature of their production areas and smaller number but greater processing (throughput) capacity of their mills. The industries in these countries are looking at creating 'macro-business' organizations made up of associations of producers, processors/packers and marketers. These aim to create an integrated supply system from production to trading, and could help in adding value to products and avoiding costly supply fragmentation (Olivae, 2000).

5.2.2. Italian Market

Italy is the second world producer and the first consumer of olive oil. The national production estimated by "Istituto di Servizi per il Mercato Agricolo Alimentare" (ISMEA) with the producers' contribution is 416.000 tons of olive oil for 2013 crop year.

Olive trees characterise Italian countryside being cultivated in 18 regions out of 20. Olive oil plays an important economic role in the economy of entire regions, in terms of employment, soil and environment protection, especially in the South where the most production is located. Puglia represents about 39% of national production followed by Calabria, 25% and Sicily, 9%. In Table 5.2 below are given some data on the olive production in Italy in the last years.

Table 5.2: Olive production in Italy, 2000-2013

	2000	2002	2005	2008	2010	2012	2013
Production							
(1000 tons)	735	656,7	879	510	430	399,2	415,5

Source: http://www.internationaloliveoil.org/, 2014

Italy is also a very large olive oil consumer. The consumption of olive oil most of the years is higher that the production. This gap is filled by imports of the products and also by gathered products from the past years. The Figure 5.9 below shows some trends in production/consumption of olive oil in Italy.

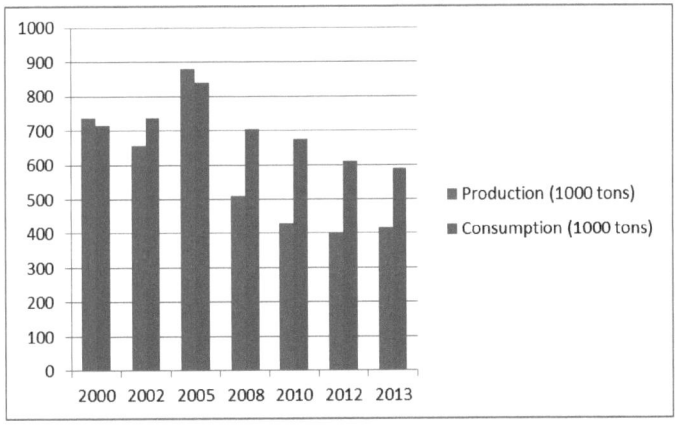

Figure 5.9: Olive oil consumption and production in Italy
Source: http://www.internationaloliveoil.org/, 2014

Referring to market in general the per capita consumption a year of olive oil is 11,9 kilogram in Italy. About 89.000 tons are imported in Italy to feed the inner market (average for 2000-2013). The import-export data for the last years are given in the Figure 5.10.

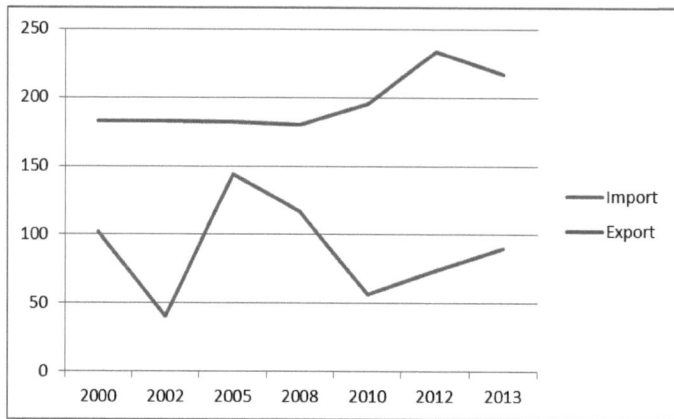

Figure 5.10: Import-export situation of olive oil in Italy, 2000-2013 (1000 tons)
Source: http://www.internationaloliveoil.org/, 2014

The Italian olive oil industry is at a competitive disadvantage in relation to the Spanish and Greek industries. They have identified the need for better organization of production and aggregation of supplies for processing in order to achieve their objectives of producing high-quality oils at competitive costs.

5.2.3. Greek Market

Greece is the third country in olive oil production. In the 9 Administrative Districts in which Greece is divided, extensive cultivation is present in the Peloponnesus (31%), Crete (20,9%) and Sterea & Evia (20,4%). Olive groves represent 20,5% of total farmland and olive oil production 14% of total plant production. In total, approximately 920.000 hectares and in average, production olive oil comes to over 358.000 tones in 2012/2013 (FAO, 2014).

In Greece, olives were harvested from November to the end of March. Farmers typically owned presses and managed the oil extraction process. A farm which produced 1.000 tones of oil per year would be considered large. Farmers generally avoided long-term production/pricing contracts with packaging companies. The custom for most farmers was to promote self-consumption sales while speculating on future prices with packaging companies. Some farmers would hold back inventories for several years in order to speculate on prices. (Even when properly stored, olive oil begins to deteriorate in quality after about one year.) In Table 5.3 are given some data on the olive production in Greece in the last years

Table 5.3: Olive production in Greece, 2000-2013

	2000	2002	2005	2008	2010	2012	2013
Production (1000 tons)	420	358,3	435	327,2	320	294,6	357,9

Source: http://www.internationaloliveoil.org/, 2014

After virgin olive oils had been received, packaging companies would clean and purify the supplies, grade and blend the oils and run a bottling operation. Greeks are also good consumers of olive oil. Interesting is the fact that they only consume their olive oil. Imports for 2000-2013 are 0 tons so the total olive oil consumption is based in domestic production.

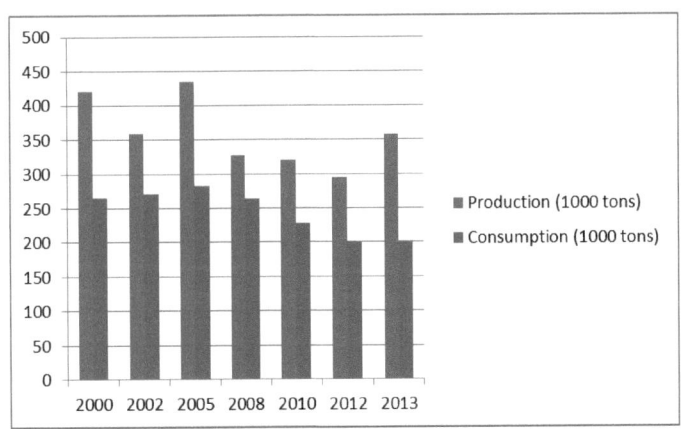

Figure 5.11: Olive oil consumption and production in Greece
Source: http://www.internationaloliveoil.org/, 2014

The Greeks pride themselves as olive oil connoisseurs and would not buy what was perceived to be lower quality Italian, Spanish or Tunisian olive oils. Although Greek olive oil had a superior reputation for quality, Greek olive oil is about 10% more expensive than Italian or Spanish oils. Almost all of the cost difference was made up in manufacturing costs. The largest Greek oil plant was perhaps one-third the size of Bertoli's largest plant in Italy. Greek plants were also used to produce both olive oils and seed oils whereas the large Italian and Spanish plants focused exclusively on olive oils.

As mentioned before, in the last 13 years the olive oil imports for Greece are 0. But they do some exports. In the Figure 5.12 below are given some data for the import-export situation in Greece.

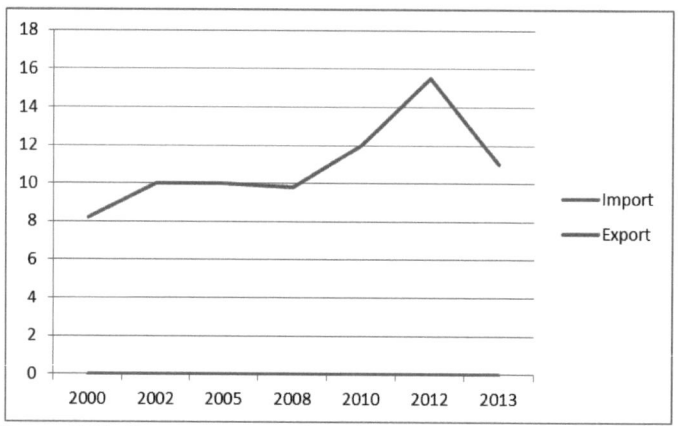

Figure 5.12: Import-export situation of olive oil in Greece, 2000-2013 (1000 tons)
Source: http://www.internationaloliveoil.org/, 2014

Three market trends were evident in the Greek olive oil industry. First, Greeks were increasingly shifting towards pre-packaged, branded oils. Urbanization, time constraints and the loosening of relationships between farmers and urban consumers were regularly cited for this trend.

Second, Greeks were moving from virgin oils to cheaper blended varieties. Increased retail stocking fees and the increased international demand for olive oil had pushed prices substantially higher. Faced with steep annual price increases, many consumers were shifting to more economical blends.

Third, the demand for bulk Greek olive oil from Spanish and Italian traders is increasing. Poor harvests in Spain combined with soaring markets in North America pushed large Spanish and Italian traders and bottlers to aggressively seek new supplies of olive oil.

A comparison of the production data of these three countries representing the top countries in olive oil production and consumption indicates the following situation shown in Figure 5.13. The Spanish superiority in olive oil production is very much evident.

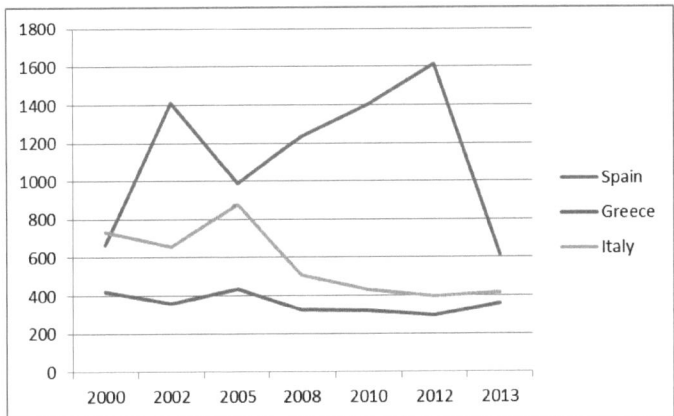

Figure 5.13: Olive production in Spain, Italy and Greece, 2000-2013
Source: http://www.internationaloliveoil.org/, 2014

6. Conclusions and Recommendations

Agriculture has been and is still the most important sector in the Albanian economy; accounting for 21% of the total GDP in 2013 and 60% of the total labor force in this sector. Even though the olive production does not take a large share in the total agricultural production, it is an industry with huge potentials that has been steadily growing during the years.

Actually Albania is importing most part of the food fats needed for the domestic trade, even though it has considerable potentials. The surface plant with olives is 42 thousand hectares, with a total number of olive trees of 3,6 million. Because of the insufficient services towards the olive tree, its production has low growth rates and with a very high yield fluctuation if we compare it with the neighbouring countries. This situation affects the olive oil processing industry resulting in small quantity and low quality olive oil.

The major activity of olive oil manufacturing industry is processing the olive of the olive producer farmers into oil against payment. This might be a reason for a minimum marketing used in this industry. From 80 manufactures that were part of this study, only 23 of them sells their product in the domestic market. Marketing elements like, packaging, guaranty, mark and name of the product, are very rare. For a long term development of the actual processing industries investments towards the names and the marks of the products are necessary.

The relationship between the farmers and processors has a big importance in the olive oil processing industry. The olive processors depend on the farmers as far as the quantity and the quality of the raw material (olives) is concerned. Olive oil producers should have long term contracts with the farmers that provide the raw material. They should offer them technical assistance, and financial support if needed to create a stable relationship with them. The farmers should feel that their product (olives) has a secured trade in the future and with stable prices.

The results of this study and their corresponding implications can be summarized as follows:

1. According to the data taken in 80 olive oil processing plants for the production year 2012-2013, the results indicated that olive oil production is privately profitable. The Private Cost Ratio has been measured and the coefficient was evaluated at 0,63, meaning that this production is profitable for the private enterprises.

2. The results of DRC analysis indicate that Albania has a comparative advantage in olive oil producing industry in the production year 2012-2013. The calculations resulted in DRC = 0,83. This means, it is socially desirable to produce and expand olive oil production in Albania, as the use of domestic factors is efficient in economic terms.

3. A sensitivity analysis was done to see how the DRC ratio reacts to different changes in different parameters of the model. Changes were done in the parameters like world market prices of olives and olive oil, exchange rate and labor force price. As a result of the sensitivity analysis it was seen that the olive oil production is very sensitive to changes in input (olives) and output (olive oil) prices of the world market and also to changes in the exchange rate. But it is not sensitive towards changes in the labor force prices. The conducted sensitivity analysis indicates that the comparative advantage of olive oil production is very fragile and could be lost easily.

4. The major determinants of the Albanian olive oil comparative advantage are the favorable world price of olive oil, the exchange rate and the price of olives as input factors for the olive oil manufacturing. The explored values of private profitability and the DRC suggest that that olive oil production is privately and socially profitable, however two important conclusions are to be emphasized particularly: firstly, the private profitability is higher than the social profitability, and secondly, social profitability is largely depended on the situation at the international market.

Olive oil production can become a very important aspect in the Albanian agriculture economy. Due to the favourable climatic conditions the main input, olives, can be cultivated in a more intensive form, despite the fact, that the areas under olive cultivation in Albania compared with the areas in Greece or Italy are very insignificant.

If under the current extensive, inefficient conditions in which the olive culture is cultivated in Albania there is however a comparative advantage (although low), this can be improved if the olive culture is cultivated more intensively. If the farmers are sure that the processing industry will act as a reliable market for their products, they will increase the production. On the other hand the increased olive cultivation will provide more raw materials for the processing industry, assuring its functioning with full capacity. The better utilisation of existing capacities in processing industry will allow favouring from the low of economies of scale and at the end effect result in lower production cost.

The government policy makers should try to implement policies encouraging the production of olive, thus putting the above described favourably accelerating development into motion. Having a more competitive and efficient product at the end will help the overall country's economy and will decrease the trade deficit of Albania.

References:

Adelman, I and Taylor, J. E., (1990), "Changing comparative advantage in food and agriculture: lessons from Mexico", Development Centre studies}/ Organisation for Economic Co-operation and Development, Paris, France

Agolli, Sh., Cipi, A., and Mance, M., (2000), "Policy Analysis Matrix: Evaluation of Comparative Advantage in the Greenhouse and Field Production", IFDC, Tirana, Albania

Agolli, Sh., Velica. R., and Mance, M., (2002), "The Olive Oil Industry: Marketing, Technology and its Competitiveness", IFDC, Tirana, Albania

Balassa, B., (1974), "New approaches to the estimation of shadow exchange rate: A comment", Oxford Economic papers, 26(2), July 1974

Bhagwati, J. N., (1998) Lectures on International Trade/ Jagdish N. Bhagwati, 2.ed. Cambridge, Mass.: MIT Press

Cardwell, M. N., (2003), "Agriculture and international trade: {law, policy and the WTO}", Wallingford

Chungsoo, K., (1983), "Evolution of Comparative Advantage: the factor proportions theory in a dynamic perspective", Tübingen, Germany

Civici, A., (2003), "The Situation and Competitiveness Level of the Agro-food Sector in Albania", Tirana, Albania

Cungu, A.,and Swinnen, J.F. M., (1998), "Albanian's Radical Agrarian Reform", Policy Research Group, Working Paper No. 15, April 1998

Edward, E. L., (1984), "Sources of International Comparative Advantage: theory and evidence", Cambridge

Ethier, W.J., (1988), "Modern International Economics", University of Pennsylvania, USA

Felderer, B., (2001), "Growth and trade in the international economy: papers for the conference "Dynamics in economic growth and international trade", Vienna, 22 - 23 June 2001

Gandolfo, G., (1987), "International Economics I: The pure theory of International Trade", University of Rome "La Sapienza", Rome, Italy

Gray, H. P., (1987), "International economic problems and policies", New York: St. Martin's Pr., XIV, 486 S

Greenaway, D., (1988), "Economic Development and International Trade", St. Martin's Press, New York, USA

Huang, J., Song, J., and Fuglie, K., (2002), "Competitiveness of Sweet Potato as animal feed in China", CCAP, Working Paper-E12

International Olive Council, (2014) http://www.internationaloliveoil.org/

Kabursi, A. A., "Lebanon's Agricultural Potential: a Policy Analysis Approach", McMaster University and Econometric Research

Kallio, P. K. S., "Old and new challenges in international agricultural trade", European review of agricultural economics; 29, 1 : Special section, Oxford : Oxford Univ. Press

Kapaj. A. (2010), "Assesing the Comprative Advantage of Albanian Olive Oil Production", IFAMR International Food &Agribusiness Management Association, Volume 13, issue 1, 2010,

KAPAJ A. (coauthor), "Olive Oil, Constituents, Quality, Health Properties and Bioconversions", INTECH-open, December 2011, ISBN 978-953-307-921-9, http://www.intechopen.com/books/olive-oil-constituents-quality-health-properties-and-bioconversions

KAPAJ A. (2013), "Agricultural Markets in a Transitioning Economy: An Albanian Case Study", CABI Publishing

KAPAJ A. "Investiogation of production oportunities and resource use efficiency in agricultural production in Albania", Anual Conference of IAMA (International Food &Agribusiness Management Association), 20th Anual World Symposium, Boston, USA, June 20, 2010

Kemp, M. C., (2001), "International trade and national welfare", Routledge frontiers of political economy, London

Keuschnigg, M., (1999), "Comparative advantage in international trade: theory and evidence", Studies in empirical economics, Florenz, European Univ. Inst., Diss.

Khachatryan, N., (2002), "Assessing the market potential of brandy produced in Armenia", University of Hohenheim, Stuttgart, Germany

Krogman, P.R. and Obstfeld, M., (2003), "International Economics: Theory and Policy", World student series, USA

Marjit, S. and Acharyya, R., (2003), "International trade, wage inequality, and the developing economy: a general equilibrium approach", Heidelberg

Monke, E. A., and Pearson, S. R. (1989), "The policy analysis matrix for agricultural development", Ithaca NY: Cornell University Press

Nguyen, M. H. and Heidhues, F., (2004), "Comparative Advantage of Vietnam's Rice Sector under Different Liberalization Scenarios", University of Hohenheim, Stuttgart, Germany

Nguyen, M. H., (2002), "Changing Comparative Advantage of Rice Production under Transformation and Trade Liberalization: a policy analysis matrix study of Vietnam's rice sector", University of Hohenheim, Stuttgart, Germany

Osmani, R., (2000) "The manual for olive cultivation", Tirana, Albania

Pearson, S., Gotsch, C. and Bahri, S., (2003), "Application of the Policy Analysis Matrix in Indonesian Agriculture", May 2003.

Robson, C., (1993), "Real world research; A resource for social scientists and practitioners researcher", Blackwell, Oxford

Samuelson, P. A., (1952), "Spatial price equilibrium and linear programming; an American economic review", vol. 42, pp. 283-303

Sánchez, J. ed., (2002), "Olive oil", European journal of lipid science and technology; 104, 9/10: Special issue, Weinheim: Wiley-VCH

Takayama, T. and Judge, G. G., (1971), "Spatial and temporal price and allocation models/ Takashi Takayama", George G. Judge, Amsterdam: North-Holland Publ. Comp.

Themelko, H., (2001), "Olive Situation in Albania and Measures for Increasing the Olive and Olive Oil Production", Centre for Rural Studies, Tirana, Albania

Tracy, M., (ed), (1998), "CAP reform: the Southern products: wine, olive oil, fruit and vegetables, bananas, cotton, tobacco", Papers by Southern European Experts, Agricultural Policy Studies

Van Marrewijk, C., (2002), "International trade and the world economy", Oxford: Oxford University Press

Vossen, P., (2000), "Olive Oil Production in Italy", Technical report on the olive oil production tour

Vossen, P., (1997), "Olive Oil Production in Spain", Technical report on the olive oil production tour

Vossen, P., (1999), "Olive Oil Technology influences on Oil quality"

Zaloshnja, E., (1997), "Analysis of Agricultural Production in Albania: Prospects for Policy Improvements", Dissertation work, Blocksburg, Virginia, USA

MoAFCP, (Ministry of Agriculture, Food and Consumers Protection), 2010, "Annual Report 2009", Tirana, Albania

MoAFCP, (Ministry of Agriculture, Food and Consumers Protection), 2013, "Annual Report 2012", Tirana, Albania

MoAFCP, (Ministry of Agriculture, Food and Consumers Protection), 2014, "Annual Statistics 2013", Tirana, Albania

MOF, (Ministry of Finance), 2013, "Annual Report", Tirana, Albania

INSTAT, (Institute of Statistics), 2012, "Statistical Yearbook 2011", Tirana, Albania

INSTAT, (Institute of Statistics), 2014, "Statistical Yearbook 2013", Tirana, Albania

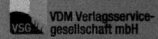